ISBN 978-0-265-93191-2
PIBN 10911733

# MAGNETIC OBSERVATIONS

AT THE

# UNITED STATES NAVAL OBSERVATORY

## 1888 AND 1889

BY

## ENSIGN J. A. HOOGEWERFF, U. S. NAVY

### Captain F. V. McNAIR, U. S. Navy
SUPERINTENDENT

PUBLICATION APPROVED BY THE CHIEF OF BUREAU OF EQUIPMENT
NAVY DEPARTMENT

WASHINGTON
GOVERNMENT PRINTING OFFICE
1890

# TABLE OF CONTENTS.

---

# I.

## BUILDINGS.

The magnetic instruments in use at the Naval Observatory are set up in two buildings, which were erected for the purpose by the Bureau of Navigation, Navy Department, in 1886–'87, and which were assigned to the Observatory in July, 1887.

These buildings are 100 feet west of the main Observatory building and are 50 feet apart on a north and south line.

The northern building, which is known as No. 1, is a one-story wooden structure, 37 feet 6 inches by 21 feet 6 inches, the length being north and south. In this building are set up the instruments for absolute observations.

The southern building, known as No. 2, is 25 feet 7 inches by 19 feet 7 inches, and consists of an underground basement, with a single-story building of wood above it. The basement is of brick, with cemented floor, the ground underneath which, and around the walls being thoroughly drained, so that the interior is kept very dry. Inside the basement and 22 inches from its walls is a wooden cell of double thickness and with a double ceiling, the upper planking of which forms the floor of the building above.

On three sides the space between the brick walls and inner cell is carried up, with double thickness of wood, about 2 feet, forming a transom or bench around the sides of the building above and giving room for six small windows, with solid wooden shutters on the outer sides, just above the level of the surrounding ground. On the fourth side is a stair-way, opening into the room above. There are six windows and a door in the wooden cell, which open into the space between it and the walls. All of these windows are kept closed to exclude dampness and to keep the temperature as nearly constant as possible. Inside the inner cell of this basement are set up the self-registering magnetographs, while a portion of the space between it and the walls is utilized for a seismoscope and seismograph. There is a copper pipe, with branches opening over each of the lights used for the magnetographs, which carries off the hot air from them into the outside air and which in that way helps to ventilate the room. The wooden building over this basement consists of an office room and a dark room for photography, etc.

In building both buildings, Nos. 1 and 2, great care was taken to exclude all magnetic material, the fastenings, etc., being entirely of copper, brass, or wood; no iron being used in any form.

The stoves in the wooden buildings are of soap-stone, with copper pipes, and wood only is burned in them. There is no provision made for heating or cooling the basement cell.

# II.

## INSTRUMENTS.

In building No. 1 are five piers, which are based on concrete and separated entirely from the floor of the building. Three of these piers are in a north and south (magnetic) line, near the west wall of the building. The northern and southern piers are at equal distances from the middle one. The southern pier is of granite, and on it is set up a large declinometer. On the middle pier, which is of brick with granite slab, is a large theodolite. On the northern pier is an arrangement for testing compasses.

*The declinometer* is a magnet 12 inches by $1\frac{1}{4}$ by 0.2 inches, hung inside a box of brass and glass by a silk suspension from a torsion circle on top of a large glass tube 60 inches high. The suspension is fastened to the magnet by a stud, on a German-silver strap, which encircles the magnet. The strap is provided with a stud on the opposite side, so that the magnet can be turned over. The magnet is fitted with a sliding weight, counterpoise, and levels. On the north end of the magnet is a small plane mirror, which can be accurately adjusted. The magnet is surrounded by a pure copper damper. A detorsion bar is also provided. The declinometer was made by FAUTH & Co., of Washington, with great care.

*The theodolite* on the middle pier has a horizontal circle of 10 inches diameter, and can be read by means of two micrometers to seconds  It is mounted on a block, which travels on rails on a second block secured to the pier-cap. The block carrying the theodolite can be moved in an east and west direction by means of a large screw. On the south side of the pier, just above the floor, is a curved scale, the radius of which is the distance from the declinometer to the scale. It is provided with adjusting screws and levels. On the block carrying the theodolite is a frame so arranged that a plumb-bob can be hung just in front of the object glass of the telescope. The point of suspension of the plumb-bob is capable of a small movement by means of a screw.

*The compass testing stand* is on the north pier on a block arranged similarly to that of the theodolite, so as to be capable of motion in an east and west direction. On the movable block are screwed two Y's, which hold the compass to be tested. Over the center of the compass are two prisms attached to the arm of an upright. These prisms are capable of adjustment, so that the opposite marks of the compass card in line with the theodolite can be seen when the theodolite is directed on the prisms.

The two other piers in this building are of oak and are in the northeast part of the building, well removed from the declinometer.

On one of these piers is set up a portable magnetometer, by ELLIOTT BROS., London, and which is their usual pattern. On the other pier is a dip circle made by KENT, of Dover.

*Magnetographs.*—In the basement of building No. 2 is a set of KEW self-recording magnetographs, by ADIE  These instruments are set on granite piers and, besides the self-recording arrangement, have telescopes and scales set on stone piers, by means of which the positions of the magnets can be read at any time.

*A seismoscope and seismograph*, the former of the MENDENHALL pattern, and the latter of the Lick Observatory pattern, are set on the cement floor (which is very solid) in the space between the basement wall and inner cell. A second seismoscope is set up on a brick pier in a small building, which is at some distance from the magnetograph building. Both seismoscopes are electrically connected to clocks in the upper part of building 2, in such a way that when they fall the corresponding clock is stopped.

# III.

## OBSERVATIONS.

*Absolute.*—The absolute observations taken are for declination, horizontal force, and inclination (dip).

*Declination.*—The declination is determined by the old method of GAUSS, as follows: In the first place, the errors of collimation, etc., are found for all the different focii used, and all observations are made with the telescope in its usual position and also inverted in the Y's, so as to eliminate all errors of the instrument as much as possible.

The reading of the horizontal circle, when the line of sight is in the meridian, is found from star observations, which are taken whenever there is reason to believe that the instrument has altered its position. When an azimuth is taken the azimuth of a fixed mark (which is a collimating telescope on a pier near the declinometer) is observed and used as a check on the position of the theodolite. The telescope is then directed on the suspension thread of the magnet and the horizontal circle clamped and read. The plumb-bob is then adjusted so as to be in the optical axis of the telescope. (The vertical wire of the telescope is so placed that there is no error of collimation when focused on the suspension thread.) The telescope is then directed on the mirror on the end of the magnet, the horizontal circle remaining clamped, and focused so that the reflected scale can be read. When the reflection of the plumb-line where it crosses the scale is seen near the vertical wire the reading of the scale is noted, and serves as the o mark of the scale. A time is selected when there is little or no change in declination and no disturbances are in progress, and these readings taken with the telescope direct and reversed and with the magnet in its usual position and reversed, and from these is found the error of the magnet for inversion and also the collimation error of the telescope for that focus. These corrections are determined at intervals of never more than three months.

After determining all corrections the telescope is directed on the suspension and the horizontal circle clamped. The telescope is then lowered on to the mirror and clamped there, and focused till the scale can be read. The instrument is not disturbed in any way now until it is thought necessary to get a new set of determinations. The reading of the reflected scale where it crosses the vertical wire of the telescope is taken twice a day and also the time, both of which are recorded in a book kept for the purpose. From these observations the declination is found at two moments of each day.

*Horizontal force.*—Previous to April, 1890, it had been the practice to make a single observation of horizontal force on Tuesday of each week, but since that time observations have been made on two successive days, at intervals of two weeks. In the vibration experiment every ninth passage of the needle is recorded up to and including the ninety-ninth, after which the observer waits for the one hundred and eightieth passage and records each ninth passage to the two hundred and seventy-ninth. The difference of the times of the passages differing 180 from each other are taken, and the mean of these divided by 180 is taken as the time of vibration. This is really the mean of 2,160 vibrations. Immediately after observing the vibrations the observation to determine the effect of the torsion is taken, the torsion circle being turned through $+90°$, $-90°$, $+360°$, and $-360°$, the mean position of the magnet being noted when the torsion circle reads 0, $+90°$, 0, $-90°$, 0, $+360°$, 0, $-360°$, and 0. From these observations the effect of torsion for $90°$ is computed. In the deflection experiment the method of sines is used, and the deflecting magnet is placed successively at distances of .3 and .4 meter from the deflected magnet on both sides, and with its north pole towards and from the deflected magnet in each position. The same mark of the scale (reflected by the mirror on the deflected magnet) is brought to the wire in the telescope by turning the horizontal circle and both verniers read and recorded for each position. The temperature of the magnet and of the bar is observed and recorded during both observations. At least once each month the vibration experiment is made with the inertia cylinder attached.

*Dip.*—It has been the practice to make observations with the dip circle between 9 and 11 a. m. and 2 and 4 p. m. on Mondays and Fridays of each week; but since April, 1890, an observation for dip has been taken immediately before and after each observation for horizontal force. Three needles are used in rotation.

*Magnetographs.*—A continuous photographic record is kept of the changes of position of the magnetograph magnets, except for a few minutes each day, between 12.20 and 12.30 p. m., when the papers are changed or slits moved. Usually two days' records are kept on each sheet. The papers are developed every few days and the ordinates measured for each hour and recorded. The temperature of the magnets is taken and recorded every three hours. Eye observations are taken and the temperatures recorded whenever an observation of horizontal force or dip is made.

At least once in three months observations are taken to determine the change in force, represented by a change of $1^{mm}$ on paper and of 1 scale division of the eye-reading scales for the horizontal and vertical force magnetographs.

The temperature corrections for the horizontal and vertical force magnetographs are also determined at least once a quarter by alternately heating and cooling the magnetograph-room and recording the readings and temperatures at frequent intervals.

*Seismoscopes.*—The clocks connected with the seismoscopes are frequently compared and their errors and rates recorded, and the seismoscopes are examined and dropped occasionally to see that they are in order.

*Seismograph.*—This instrument is examined from time to time and kept in good working order.

Up to the present time there has been no earthquake shock since setting up these instruments.

*Occasional observations.*—Occasional observations are made when possible with different instruments, in order to check the regular observations, and the temperature co-efficients, etc, furnished at Kew.

# IV.

## REDUCTION OF OBSERVATIONS. ·

*Declination.*—The ordinate on the photographic trace of declination corresponding to the time when the absolute declination is taken, is measured with a scale whose divisions are equal to minutes of arc of the movement of the magnet. This ordinate subtracted from the true declination gives the value of the base-line. Two values are thus obtained for each day, and the mean of these for the month is taken as the base-line value for that month. The traces for each month are then copied with a pantograph, so that the base-lines correspond, and in this way a composite curve of the declination traces for the month is formed. The mean curve is then drawn by the eye on a sheet of tracing-paper laid over the composite curve, and the ordinates measured for each hour with the scale graduated to minutes explained above. These measurements added to the value of the base-line are taken as the value of the declination for the corresponding hour during the month. The mean of the twenty-four hours is taken as the mean declination for the month. By this method all disturbances are practically eliminated in finding the mean value of the declination. All traces on the composite curve which differ at any hour 2 minutes or more from the mean are measured and recorded as disturbances.

*Horizontal force.*—The ordinates of the photographic traces for each day are measured at each hour and recorded. The proper temperature correction is then applied to each ordinate to reduce all the ordinates for a month to the same temperature. The ordinates (reduced to this temperature) corresponding to the times when the absolute horizontal intensity was observed during the month are then taken and placed opposite their corresponding absolute force. From them is deduced the ordinate which at about the middle of the month corresponds to a given force. From this as a base, with the change in ordinate corresponding to change in force, is constructed a table for the month, and with it all the ordinates (corrected for temperature) are converted into terms of absolute force. The change in ordinate from the middle of one month to the middle of the next month, due to loss of magnetism of the magnet, is then found, and the proper correction for it applied to all the results.

In the horizontal-force traces increase of ordinate denotes increase of force, and the mean value for the year 1889 is $1^{mm}$ of ordinate $= .000048$ C. G. S. unit.

Increase in temperature caused a decrease in ordinate, and the mean value of the temperature correction for 1889 is $1° C. = 1.24^{mm}$ change of ordinate.

*Vertical force.*—The ordinates of the vertical-force traces are measured, recorded, and corrected for temperature in the same manner as those of the horizontal-force traces. The absolute vertical force and the corresponding ordinate are then found, the former from the dip observations and the horizontal force at the middle of the corresponding dip observation, and the latter being the ordinate corresponding to the middle time of the dip observation reduced to the temperature to which all the ordinates for the month are reduced. The conversion into absolute force is then done in a similar manner to that used for the horizontal-force ordinates, and the correction for loss of magnetism found and applied in a similar way also.

In the vertical-force traces increase of ordinate denotes decreasing force, and the values for 1889 are as follows:

From January to August 10, .1$^{mm}$ of ordinate     $=.000048$ C. G. S. unit.

From August 10 to December 31, 1$^{mm}$ of ordinate $=.000301$ C. G. S. unit.

Increase in temperature caused an increase in ordinate, and the temperature corrections for 1889 are as follows:

From January to August 10, 1° C.     $= 1.60^{mm}$ change of ordinate.

From August 10 to December 31, 1° C. $=0.27^{mm}$ change of ordinate.

The loss of sensibility caused by the magnet having to be re-adjusted makes the records for the latter part of the year much less reliable than those for the first part.

*Total force.*—The total force is calculated from the horizontal and vertical forces for the same time.

*Dip.*—The dip is calculated from the horizontal and vertical forces for the same time.

## V.

## PERSONNEL OF THE OFFICE.

Lieut. H. W. SCHAEFER, U. S. Navy, was in charge of the buildings from October 25, 1886, to December 31, 1887, and superintended the arrangements of the buildings and the setting up of the piers and some of the instruments.

Lieut. WILLIAM P. ELLIOTT, U. S. Navy, was ordered as assistant December 18, 1886, and succeeded Lieutenant SCHAEFER, in charge, December 31, 1887. Lieutenant ELLIOTT had charge until March 15, 1889. He completed the setting up of the instruments and started the taking of the observations.

Ensign C. C. MARSH, U. S. Navy, was ordered as assistant December 19, 1887, and was in charge from March 15, 1889, to January 6, 1890. Under Ensign MARSH the observations were regularly taken and recorded, and he devised the composite curves of declination.

Ensign J. A. HOOGEWERFF, U. S. Navy, was ordered as assistant June 1, 1889, and succeeded Mr. MARSH January 6, 1890.

Ensign W. B. HOGGATT, U. S. Navy, was ordered as assistant January 11, 1890.

The absolute observations were reduced by Ensigns MARSH and HOOGEWERFF in December, 1889, and the reduction of the traces and the completion of the work was done by Ensign HOOGEWERFF, who was ably assisted in this long and tedious work by Ensign HOGGATT.

# VI.

## RESULTS.

All dates in the tables are civil dates, and the hours are for the seventy-fifth meridian (west of Greenwich) mean time, which is 8 minutes 12.09 seconds fast of Washington meridian time.

Table   I. Mean hourly values of the declination for 1888 and 1889, and for the other magnetic elements for 1889, compiled from Tables II, III, and IV. In the horizontal-force and vertical-force results January, having only a few days, was thrown out.

      II. Mean hourly declination for each month of 1888 and 1889, taken from monthly composite curves.

     III. Mean hourly values of the horizontal force for each month of 1889, in C. G. S. units (dynes).

    IV. Mean hourly values of the vertical force for each month of 1889, in C. G. S. units (dynes).

     V. Declination ordinates for each hour, expressed in minutes of arc, taken from daily declination traces.

    VI. Hourly values of the horizontal force, being the ordinates taken from the daily traces, corrected for temperature and converted into absolute measure with all corrections, except that the disturbances are not eliminated. See "Horizontal force."

   VII. Same as Table VI, except that for horizontal force read vertical force. See "Vertical force."

  VIII Summary of disturbances in declination during 1888 and 1889 which differed 2 minutes or more from the mean monthly curve, as determined by the composite curves of declination.

    IX. Observations for horizontal intensity, 1888 and 1889.

    X. Observations for inclination (dip), 1888 and 1889.

# VII.

## PLATES.

All dates given, except in Plate II, are astronomical dates (in which the day begins twelve hours later than the civil day). This is done because the magnetographs are set to run from noon to noon.

The hours are for the seventy-fifth meridian (west of Greenwich) mean time, which is 8 minutes 12.09 seconds fast of Washington meridian time.

Plate   I. Examples of the daily photographic traces of declination, horizontal and vertical force. Two traces are on each sheet to economize paper. The days are selected as showing the difference between disturbed and undisturbed days.

Plate   II. Mean diurnal variation of the magnetic elements for the year 1889, with
            the declination for 1888 added.   In this plate the civil date is used.
        III. Monthly composite curves of declination for first six months of 1888.
        IV. Monthly composite curves of declination for last six months of 1888.
        V. Monthly composite curves of declination for first six months of 1889.
        VI. Monthly composite curves of declination for last six months of 1889.
        VII to XIV. Comparisons of disturbed days of declination at the Naval Observ-
            atory, Washington, D. C.; at the U. S. Coast and Geodetic Survey
            Magnetic Observatory at Los Angeles, Cal.; at the Canadian Magnetic
            Observatory at Toronto, Canada; and at the Observatoire Physique
            Central de Russie at Pawlowsk, Russia.

The traces are all placed for the same time, which is seventy-fifth meridian (west
of Greenwich) mean time, and were reduced to the same length of base-line by means
of a pantograph.   The traces, or copies of the traces, from Los Angeles, Toronto, and
Pawlowsk were obtained through the kindness of the Superintendent of the Coast and
Geodetic Survey, the Director of the Meteorological Service of Canada, Prof. Charles
Carpmael, and the Director of the Observatoire Physique Central de Russie, Dr. H.
Wild, to whom the thanks of this Observatory are tendered.

TABLE I.—*Mean hourly values of the magnetic elements for the year.*

| Hours. | 1888. Declination (westerly). | 1889. Declination (westerly). | Inclination (dip). | Horizontal force (C. G. S. units). | Vertical force (C. G. S. units). | Total force (C. G. S. units). |
|---|---|---|---|---|---|---|
| A. M. | o ′ ″ | o ′ ″ | o ′ ″ | | | |
| 1 | 3 58 28 | 4 01 17 | 71 05 50 | .198717 | .580314 | .613395 |
| 2 | 3 58 31 | 4 01 17 | 05 45 | 731 | 309 | 395 |
| 3 | 3 58 19 | 4 01 08 | 05 48 | 722 | 308 | 391 |
| 4 | 3 59 01 | 4 00 51 | 05 47 | 726 | 313 | 396 |
| 5 | 3 57 47 | 4 00 32 | 05 45 | 737 | 328 | 414 |
| 6 | 3 57 06 | 4 00 03 | 05 44 | 746 | 341 | 430 |
| 7 | 3 56 16 | 3 59 11 | 05 49 | 730 | 340 | 423 |
| 8 | 3 55 46 | 3 58 49 | 06 05 | 677 | 334 | 407 |
| 9 | 3 56 23 | 3 59 20 | 06 26 | 605 | 312 | 356 |
| 10 | 3 58 03 | 4 00 45 | 06 38 | 554 | 275 | 305 |
| 11 | 4 00 29 | 4 02 59 | 06 32 | 558 | 237 | 270 |
| 12 | 4 02 07 | 4 04 15 | 06 20 | 597 | 239 | 285 |
| P. M. | | | | | | |
| 1 | 4 02 11 | 4 04 51 | 06 08 | 658 | 307 | 368 |
| 2 | 4 02 02 | 4 04 43 | 05 55 | 709 | 332 | 410 |
| 3 | 4 01 22 | 4 04 02 | 05 49 | 734 | 355 | 439 |
| 4 | 4 00 13 | 4 02 54 | 05 50 | 740 | 376 | 461 |
| 5 | 3 59 14 | 4 01 54 | 05 52 | 732 | 378 | 460 |
| 6 | 3 58 38 | 4 01 25 | 05 59 | 711 | 378 | 453 |
| 7 | 3 58 27 | 4 01 14 | 05 58 | 710 | 370 | 445 |
| 8 | 3 58 16 | 4 01 04 | 06 00 | 704 | 369 | 442 |
| 9 | 3 58 16 | 4 00 58 | 05 58 | 710 | 365 | 440 |
| 10 | 3 58 14 | 4 00 52 | 05 59 | 704 | 360 | 434 |
| 11 | 3 58 19 | 4 00 58 | 05 56 | 709 | 348 | 424 |
| 12 | 3 58 19 | 4 01 01 | 05 53 | 713 | 329 | 407 |
| Mean .. | 3 58 47 | 4 01 31 | 71 05 59 | .198693 | .580330 | .613402 |

TABLE II.—*Mean hourly declination for each month.  Being the*

The sum of the base-line value for the month and the numbers given in the table

| Months. | A. M. | | | | | | | | | | | |
|---|---|---|---|---|---|---|---|---|---|---|---|---|
| | 1. | 2. | 3. | 4. | 5. | 6. | 7. | 8. | 9. | 10. | 11. | 12. |
| **1888.** | | | | | | | | | | | | |
| January | 56.85 | 57.25 | 57.10 | 57.05 | 56.80 | 57.30 | 57.50 | 56.70 | 56.20 | 54.25 | 55.80 | 57.65 |
| February | 56.50 | 56.35 | 55.95 | 55.75 | 55.65 | 55.15 | 55.20 | 54.95 | 54.40 | 55.60 | 57.65 | 59.60 |
| March | 56.85 | 56.90 | 56.80 | 56.40 | 56.35 | 56.25 | 55.45 | 54.15 | 54.65 | 55.70 | 58.00 | 61.00 |
| April | 57.05 | 57.00 | 56.85 | 56.55 | 56.30 | 55.55 | 54.80 | 55.05 | 56.05 | 58.15 | 60.70 | 62.90 |
| May | 57.30 | 57.30 | 57.50 | 56.95 | 56.45 | 55.10 | 53.60 | 53.55 | 55.45 | 58.10 | 61.00 | 62.50 |
| June | 56.85 | 56.90 | 56.70 | 56.10 | 55.70 | 53.95 | 52.50 | 52.25 | 53.75 | 56.10 | 59.10 | 61.95 |
| July | 57.00 | 57.00 | 56.90 | 56.40 | 55.80 | 54.10 | 52.50 | 52.00 | 53.20 | 56.00 | 59.30 | 61.20 |
| August | 57.70 | 57.60 | 57.00 | 56.60 | 56.30 | 54.80 | 53.00 | 52.80 | 54.00 | 57.60 | 60.70 | 62.00 |
| September | 58.00 | 57.80 | 57.60 | 57.00 | 57.00 | 56.00 | 55.00 | 54.50 | 56.00 | 59.40 | 62.80 | 63.30 |
| October | 58.10 | 58.30 | 58.10 | 58.30 | 58.00 | 57.80 | 57.10 | 56.00 | 56.60 | 58.30 | 60.50 | 61.00 |
| November | 59.60 | 60.00 | 59.40 | 59.30 | 59.30 | 59.30 | 58.90 | 58.30 | 58.10 | 58.90 | 60.20 | 61.00 |
| December | 58.40 | 58.50 | 58.70 | 58.60 | 58.60 | 58.60 | 58.30 | 57.60 | 57.00 | 57.20 | 58.90 | 60.00 |
| Year | 57.52 | 57.57 | 57.38 | 57.08 | 56.85 | 56.16 | 55.32 | 54.82 | 55.45 | 57.11 | 59.55 | 61.17 |
| **1889.** | | | | | | | | | | | | |
| January | 69.6 | 69.2 | 69.0 | 69.0 | 69.0 | 69.0 | 68.6 | 67.9 | 67.2 | 68.0 | 68.0 | 69.9 |
| February | 68.9 | 69.3 | 69.7 | 69.3 | 69.2 | 69.0 | 69.0 | 68.0 | 68.0 | 68.6 | 69.9 | 70.2 |
| March | 69.6 | 69.6 | 69.2 | 69.0 | 69.0 | 69.3 | 68.2 | 67.6 | 68.0 | 68.6 | 71.1 | 71.8 |
| April | 70.1 | 69.5 | 69.5 | 69.6 | 69.1 | 68.0 | 67.0 | 67.0 | 67.6 | 69.2 | 72.2 | 73.3 |
| May | 71.0 | 70.9 | 70.6 | 70.1 | 69.5 | 68.9 | 68.0 | 68.0 | 69.1 | 71.6 | 74.5 | 75.6 |
| June | 70.9 | 70.9 | 70.9 | 70.4 | 69.4 | 68.9 | 67.3 | 67.6 | 68.8 | 71.3 | 73.8 | 74.4 |
| July | 72.6 | 72.5 | 72.0 | 71.5 | 71.4 | 69.8 | 68.3 | 68.5 | 69.8 | 71.2 | 74.1 | 75.3 |
| August | 72.8 | 73.0 | 73.2 | 73.0 | 72.3 | 71.3 | 69.5 | 69.0 | 71.2 | 74.0 | 76.7 | 78.5 |
| September | 74.3 | 74.2 | 73.7 | 73.2 | 73.1 | 72.5 | 70.9 | 71.2 | 72.3 | 74.2 | 77.1 | 79 3 |
| October | 75.0 | 75.1 | 74.9 | 74.9 | 74.5 | 74.3 | 74.0 | 72.8 | 73.0 | 74.5 | 76.9 | 78.3 |
| November | 73.4 | 73.6 | 73.5 | 73.5 | 72.7 | 72.6 | 72.4 | 72.1 | 71.8 | 72.5 | 74.3 | 75.8 |
| December | 73.0 | 73.3 | 73.2 | 72.5 | 73.0 | 72.9 | 72.8 | 71.9 | 71.0 | 71.1 | 73.0 | 74.5 |
| Year | 71.77 | 71.76 | 71.62 | 71.33 | 71.02 | 70.54 | 69.67 | 69.30 | 69.82 | 71.23 | 73.47 | 74.74 |

*ordinates taken from the composite curve of declination for the month.*

for any hour of that month is the mean declination for that hour of the month.

| P.M. | | | | | | | | | | | | Mean. | Range. | Value of base-line. | Mean declination (west). |
|---|---|---|---|---|---|---|---|---|---|---|---|---|---|---|---|
| 1. | 2. | 3. | 4. | 5. | 6. | 7. | 8. | 9. | 10. | 11. | 12. | | | | |
| 59.35 | 59.60 | 59.10 | 58.40 | 57.15 | 56.95 | 56.70 | 56.25 | 56.20 | 56.10 | 56.50 | 56.50 | 57.05 | 5.35 | 3 01 18.0 | 3 58 21.0 |
| 59.40 | 59.35 | 58.15 | 57.15 | 56.90 | 56.40 | 56.30 | 56.05 | 55.65 | 55.85 | 55.75 | 56.05 | 56.49 | 5.20 | 3 01 18.0 | 3 57 47.4 |
| 61.05 | 61.15 | 60.95 | 60.00 | 58.45 | 57.55 | 57.20 | 56.75 | 56.80 | 56.80 | 56.85 | 56.75 | 57.45 | 7.00 | 3 00 54.0 | 3 58 21.0 |
| 61.25 | 61.15 | 61.00 | 59.80 | 58.55 | 57.65 | 57.70 | 57.75 | 57.55 | 57.30 | 57.30 | 56.95 | 57.95 | 8.10 | 3 00 42.0 | 3 58 39.0 |
| 62.70 | 62.30 | 61.25 | 60.20 | 58.75 | 57.75 | 57.65 | 57.40 | 57.20 | 57.25 | 57.35 | 57.30 | 58.00 | 9.15 | 3 00 42.0 | 3 58 42.0 |
| 61.95 | 61.25 | 60.45 | 58.85 | 57.65 | 56.70 | 56.25 | 56.15 | 56.55 | 56.70 | 56.45 | 56.55 | 56.97 | 9.70 | 3 01 00.0 | 3 57 58.2 |
| 61.00 | 61.30 | 60.60 | 59.40 | 58.40 | 57.60 | 57.20 | 57.20 | 57.10 | 57.00 | 57.00 | 56.90 | 57.17 | 9.30 | 3 01 06.0 | 3 58 16.2 |
| 62.60 | 62.70 | 62.00 | 60.00 | 58.60 | 57.90 | 57.90 | 57.90 | 57.90 | 57.60 | 57.80 | 58.00 | 57.96 | 9.90 | 3 01 02.0 | 3 58 59.6 |
| 63.00 | 62.10 | 60.90 | 59.00 | 58.40 | 58.40 | 58.40 | 58.30 | 58.40 | 58.50 | 58.30 | 58.10 | 58.59 | 8.80 | 3 01 02.0 | 3 59 37.4 |
| 61.10 | 60.70 | 59.80 | 59.00 | 58.90 | 58.80 | 58.70 | 58.50 | 58.30 | 58.00 | 58.20 | 58.00 | 58.59 | 5.10 | 3 00 48.0 | 3 59 23.4 |
| 61.30 | 61.40 | 61.00 | 60.30 | 59.80 | 58.90 | 58.60 | 58.40 | 58.60 | 58.80 | 59.00 | 59.10 | 59.48 | 3.30 | 3 00 24.0 | 3 59 52.8 |
| 60.30 | 60.10 | 59.80 | 59.10 | 58.10 | 57.80 | 57.50 | 57.30 | 57.60 | 57.60 | 58.00 | 58.20 | 58.41 | 3.30 | 3 01 00.0 | 3 59 24.6 |
| 61.25 | 61.09 | 60.42 | 59.27 | 58.30 | 57.70 | 57.51 | 57.33 | 57.32 | 57.29 | 57.37 | 57.37 | 57.84 | 6.43 | 3 00 56.3 | 3 58 46.8 |
| 70.9 | 70.6 | 70.4 | 69.6 | 68.5 | 68.5 | 68.5 | 68.0 | 68.4 | 68.0 | 68.6 | 69.0 | 68.89 | 3.70 | 2 50 59.0 | 3 59 52.4 |
| 70.7 | 70.9 | 70.8 | 70.0 | 69.5 | 69.0 | 69.0 | 68.4 | 68.5 | 68.5 | 68.5 | 68.5 | 69.24 | 2.90 | 2 51 08.0 | 4 00 22.4 |
| 73.0 | 73.0 | 72.9 | 71.9 | 70.9 | 70.5 | 70.0 | 69.9 | 69.6 | 69.4 | 69.6 | 69.6 | 70.05 | 5.40 | 2 51 31.0 | 4 01 34.0 |
| 73.9 | 74.1 | 74.0 | 73.0 | 71.9 | 70.9 | 70.2 | 69.6 | 69.1 | 69.3 | 69.9 | 69.6 | 70.32 | 7.10 | 2 51 33.0 | 4 01 52.2 |
| 75.9 | 75.0 | 74.4 | 73.4 | 72.1 | 71.0 | 71.4 | 71.3 | 71.0 | 71.0 | 70.6 | 71.2 | 71.50 | 7.90 | 2 50 38.4 | 4 02 08.4 |
| 75.2 | 75.6 | 74.6 | 73.9 | 72.6 | 71.6 | 71.1 | 71.3 | 71.1 | 70.8 | 71.1 | 70.6 | 71.42 | 8.30 | 2 49 46.9 | 4 01 12.1 |
| 76.8 | 77.0 | 76.8 | 75.0 | 73.0 | 72.5 | 72.9 | 73.0 | 72.8 | 72.2 | 71.9 | 71.5 | 72.60 | 8.70 | 2 49 15.0 | 4 01 51.0 |
| 77.8 | 77.5 | 76.1 | 74.4 | 73.2 | 72.8 | 73.0 | 73.0 | 73.2 | 72.9 | 73.0 | 72 8 | 73.51 | 9.50 | 2 48 55.0 | 4 02 25.6 |
| 80.0 | 79.2 | 77.5 | 76.0 | 75.0 | 75.0 | 74.6 | 74.6 | 74.5 | 74.5 | 74.4 | 74.4 | 74.82 | 9.10 | 2 47 44.0 | 4 02 33.2 |
| 77.9 | 78.4 | 77.0 | 75.9 | 75.0 | 74.9 | 71.6 | 74.8 | 74.5 | 74.8 | 74.5 | 74.9 | 75.22 | 5.60 | 2 46 26.0 | 4 01 39.2 |
| 76.9 | 76.5 | 75.4 | 74.3 | 74.0 | 73.3 | 73.0 | 72.6 | 72.7 | 72.6 | 72.9 | 73.0 | 73.56 | 5.10 | 2 47 23.0 | 4 00 56.6 |
| 75.0 | 74.6 | 74.2 | 73.3 | 73.0 | 72.8 | 72.4 | 72.1 | 72.0 | 72.2 | 72.4 | 72.9 | 72.88 | 4.00 | 2 48 52.0 | 4 01 44.8 |
| 75.33 | 75.20 | 74.51 | 73.39 | 72.39 | 71.90 | 71.72 | 71.55 | 71.45 | 71.35 | 71.45 | 71.50 | 72.00 | 6.03 | 2 49 30.9 | 4 01 30.9 |

TABLE III.—*Mean hourly values of the horizontal force, in C. G. S. units (dynes).*

1889.

| Hour. | January (4 days). | February. | March. | April. | May. | June. | July. | August. | September. | October. | November. | December. |
|---|---|---|---|---|---|---|---|---|---|---|---|---|
| **A. M.** | | | | | | | | | | | | |
| 1 | .198935 | .198827 | .198783 | .198778 | .198695 | .198648 | .198639 | .198805 | .198693 | .198694 | .198655 | .198675 |
| 2 | 934 | 828 | 803 | 785 | 701 | 653 | 641 | 775 | 713 | 687 | 673 | 680 |
| 3 | 932 | 849 | 785 | 779 | 697 | 656 | 656 | 777 | 697 | 696 | 658 | 689 |
| 4 | 968 | 860 | 800 | 778 | 697 | 646 | 621 | 765 | 715 | 715 | 679 | 709 |
| 5 | 958 | 862 | 811 | 794 | 713 | 670 | 627 | 778 | 718 | 719 | 701 | 719 |
| 6 | 979 | 881 | 805 | 797 | 725 | 682 | 641 | 796 | 713 | 727 | 704 | 732 |
| 7 | 975 | 898 | 807 | 757 | 694 | 644 | 617 | 765 | 696 | 693 | 709 | 747 |
| 8 | 993 | 881 | 768 | 709 | 621 | 580 | 556 | 684 | 591 | 629 | 683 | 741 |
| 9 | .199019 | 870 | 712 | 622 | 500 | 497 | 489 | 571 | 501 | 563 | 614 | 718 |
| 10 | 019 | 809 | 663 | 594 | 456 | 456 | 434 | 522 | 448 | 527 | 540 | 650 |
| 11 | .198938 | 768 | 679 | 611 | 536 | 500 | 429 | 560 | 442 | 532 | 517 | 567 |
| 12 | 878 | 732 | 670 | 670 | 639 | 576 | 478 | 656 | 536 | 553 | 510 | 548 |
| **P. M.** | | | | | | | | | | | | |
| 1 | 885 | 744 | 726 | 734 | 742 | 648 | 560 | 736 | 611 | 609 | 547 | 584 |
| 2 | 891 | 756 | 756 | 780 | 794 | 714 | 630 | 805 | 670 | 644 | 623 | 630 |
| 3 | 886 | 790 | 808 | 792 | 789 | 728 | 658 | 825 | 707 | 670 | 627 | 684 |
| 4 | 869 | 837 | 816 | 778 | 743 | 725 | 682 | 805 | 713 | 680 | 649 | 707 |
| 5 | 915 | 880 | 808 | 791 | 734 | 672 | 673 | 781 | 697 | 647 | 651 | 722 |
| 6 | 927 | 846 | 766 | 755 | 701 | 655 | 641 | 778 | 670 | 674 | 640 | 699 |
| 7 | 942 | 852 | 784 | 735 | 697 | 653 | 636 | 774 | 701 | 672 | 612 | 690 |
| 8 | 931 | 818 | 788 | 756 | 696 | 650 | 623 | 776 | 682 | 672 | 601 | 682 |
| 9 | 911 | 823 | 785 | 747 | 690 | 667 | 626 | 789 | 708 | 670 | 618 | 683 |
| 10 | 930 | 816 | 766 | 717 | 687 | 664 | 637 | 773 | 696 | 682 | 629 | 679 |
| 11 | 890 | 806 | 786 | 729 | 700 | 667 | 643 | 773 | 714 | 680 | 635 | 665 |
| 12 | 906 | 824 | 784 | 764 | 690 | 651 | 641 | 793 | 697 | 678 | 655 | 666 |
| Mean | .198934 | .198827 | .198769 | .198740 | .198678 | .198638 | .198605 | .198745 | .198655 | .198655 | .198630 | .198678 |

**TABLE IV.—*Mean hourly values of the vertical force, in C. G. S. units (dynes).***

1889.

| Hour. | January (6 days). | February. | March. | April. | May. | June. | July. | August. | September. | October. | November. | December. |
|---|---|---|---|---|---|---|---|---|---|---|---|---|
| **A. M.** | | | | | | | | | | | | |
| 1 | .581805 | .581387 | .581236 | .580896 | .580507 | .580142 | .579523 | .579803 | .579867 | .580218 | .579682 | .580191 |
| 2 | 801 | 387 | 230 | 917 | 504 | 137 | 517 | 810 | 846 | 213 | 659 | 179 |
| 3 | 800 | 388 | 228 | 920 | 512 | 131 | 496 | 830 | 853 | 203 | 642 | 180 |
| 4 | 802 | 393 | 263 | 920 | 520 | 131 | 503 | 824 | 856 | 210 | 651 | 174 |
| 5 | 797 | 394 | 246 | 927 | 532 | 141 | 539 | 899 | 854 | 221 | 667 | 190 |
| 6 | 803 | 407 | 251 | 939 | 544 | 149 | 549 | 941 | 887 | 233 | 669 | 185 |
| 7 | 802 | 406 | 256 | 937 | 548 | 150 | 551 | 939 | 887 | 240 | 659 | 171 |
| 8 | 799 | 394 | 251 | 933 | 539 | 151 | 541 | 905 | 884 | 243 | 657 | 173 |
| 9 | 774 | 370 | 238 | 911 | 519 | 136 | 524 | 864 | 882 | 235 | 619 | 131 |
| 10 | 764 | 354 | 216 | 905 | 468 | 098 | 495 | 816 | 860 | 197 | 580 | 041 |
| 11 | 731 | 324 | 177 | 828 | 441 | 078 | 467 | 768 | 809 | 152 | 539 | 028 |
| 12 | 718 | 281 | 172 | 840 | 452 | 073 | 467 | 753 | 774 | 149 | 611 | 059 |
| **P. M.** | | | | | | | | | | | | |
| 1 | 775 | 349 | 199 | 867 | 472 | 083 | 480 | 700 | 961 | 217 | 800 | 247 |
| 2 | 791 | 370 | 215 | 904 | 509 | 104 | 513 | 789 | 982 | 238 | 804 | 227 |
| 3 | 786 | 390 | 236 | 905 | 552 | 127 | 543 | 862 | .580030 | 248 | 826 | 190 |
| 4 | 797 | 391 | 244 | 934 | 566 | 178 | 564 | 938 | 059 | 278 | 791 | 190 |
| 5 | 798 | 382 | 259 | 944 | 573 | 178 | 588 | 958 | 049 | 273 | 787 | 166 |
| 6 | 798 | 365 | 275 | 942 | 564 | 171 | 591 | 936 | 051 | 275 | 809 | 174 |
| 7 | 800 | 363 | 274 | 931 | 546 | 157 | 581 | 886 | 031 | 273 | 837 | 193 |
| 8 | 803 | 380 | 283 | 928 | 540 | 147 | 571 | 877 | 003 | 277 | 853 | 205 |
| 9 | 800 | 382 | 282 | 931 | 538 | 142 | 557 | 882 | .579978 | 263 | 835 | 225 |
| 10 | 798 | 382 | 280 | 921 | 533 | 141 | 546 | 870 | 977 | 245 | 825 | 235 |
| 11 | 794 | 373 | 275 | 921 | 524 | 136 | 534 | 862 | 958 | 232 | 793 | 225 |
| 12 | 787 | 364 | 258 | 912 | 503 | 131 | 524 | 856 | 900 | 224 | 751 | 195 |
| Mean.. | .581789 | .581374 | .581243 | .580913 | .580521 | .580134 | .579532 | .579860 | .579927 | .580231 | .579723 | .580174 |

TABLE V.—*Declination,*

Ordinates, expressed in minutes of arc, taken from the daily declination traces.  The ordinate

Base-line value

| Day. | A. M. | | | | | | | | | | | |
|---|---|---|---|---|---|---|---|---|---|---|---|---|
|  | 1. | 2. | 3. | 4. | 5. | 6. | 7. | 8. | 9. | 10. | 11. | 12. |
| 1 | ... | ... | ... | ... | ... | ... | ... | ... | ... | ... | ... | ... |
| 2 | ... | ... | ... | ... | ... | ... | ... | ... | ... | ... | ... | ... |
| 3 | ... | ... | ... | ... | ... | ... | ... | ... | ... | ... | ... | ... |
| 4 | ... | ... | ... | ... | ... | ... | ... | ... | ... | ... | ... | ... |
| 5 | ... | ... | ... | ... | ... | ... | ... | ... | ... | ... | ... | ... |
| 6 | ... | ... | ... | ... | ... | ... | ... | ... | ... | ... | ... | ... |
| 7 | ... | ... | ... | ... | ... | ... | ... | ... | ... | ... | ... | ... |
| 8 | ... | ... | ... | ... | ... | ... | ... | ... | ... | ... | ... | ... |
| 9 | ... | ... | ... | ... | ... | ... | ... | ... | ... | ... | ... | ... |
| 10 | 71.0 | 70.9 | 70.9 | 71.0 | 70.6 | 71.5 | 73.8 | 68.5 | 67.1 | ... | ... | ... |
| 11 | 70.1 | 70.3 | 69.6 | 69.0 | 70.4 | 72.5 | 70.5 | 68.0 | 67.4 | ... | ... | ... |
| 12 | 71.5 | 71.5 | 71.2 | 70.3 | 74.3 | 74.0 | 71.5 | 69.8 | 68.2 | 68.0 | 70.5 | 71.8 |
| 13 | 70.4 | 69.6 | 70.0 | 69.8 | 70.4 | 71.0 | 70.4 | 69.5 | 67.7 | 67.6 | 70.7 | 70.7 |
| 14 | 70.5 | 70.6 | 71.0 | 70.4 | 70.5 | 70.3 | 70.0 | 68.8 | 67.3 | 67.0 | 69.4 | 70.0 |
| 15 | 70.7 | 70.6 | 70.6 | 70.6 | 70.6 | 70.5 | 70.5 | 69.4 | 68.3 | 68.5 | 69.4 | 71.0 |
| 16 | 70.5 | 70.6 | 70.6 | 71.0 | 70.0 | 70.0 | 69.6 | 69.0 | 67.7 | 68.0 | 69.4 | 71.5 |
| 17 | 67.5 | 67.2 | 67.5 | 67.5 | 67.5 | 67.4 | 67.5 | 65.7 | 64.0 | 63.5 | 65.5 | 68.6 |
| 18 | 67.0 | 66.9 | 67.0 | 66.0 | 66.2 | 66.4 | 66.3 | 65.0 | 64.4 | 64.3 | 65.8 | 68.7 |
| 19 | 68.5 | 68.5 | 68.5 | 68.4 | 68.3 | 68.0 | 67.6 | 66.2 | 65.3 | 66.3 | 67.9 | 70.0 |
| 20 | 63.4 | 59.5 | 63.3 | 65.6 | 64.7 | 66.5 | 65.5 | 67.3 | 71.4 | 68.7 | 69.1 | 73.0 |
| 21 | 69.5 | 67.5 | 69.6 | 68.6 | 68.9 | 67.5 | 67.0 | 65.4 | 65.3 | ... | 69.0 | 69.5 |
| 22 | 69.5 | 68.5 | 68.2 | 68.5 | 66.7 | 66.6 | 66.7 | 65.4 | 66.4 | 68.4 | 70.8 | 72.3 |
| 23 | 69.3 | 69.5 | 70.4 | 68.7 | 68.5 | 68.5 | 68.2 | 65.6 | 66.2 | 67.5 | 68.6 | 70.4 |
| 24 | 68.4 | 68.5 | 68.5 | 68.4 | 67.5 | 67.7 | 66.1 | 65.0 | 65.5 | 67.2 | 69.5 | 71.5 |
| 25 | 68.6 | 68.8 | 68.6 | 68.6 | 68.5 | 68.4 | 67.3 | 66.0 | 65.6 | 65.6 | 68.0 | 70.0 |
| 26 | 68.4 | 68.8 | 68.5 | 68.5 | 68.5 | 68.0 | 67.6 | 67.5 | 66.9 | 66.7 | 68.6 | 70.0 |
| 27 | 68.4 | 68.7 | 68.6 | 68.5 | 68.5 | 67.8 | 67.5 | 67.1 | 67.3 | 67.2 | 66.9 | ... |
| 28 | 68.4 | 68.5 | 67.3 | 67.4 | 66.5 | 67.5 | 67.4 | 67.2 | 67.5 | 65.6 | 66.6 | 69.0 |
| 29 | 68.4 | 68.2 | 68.4 | 70.5 | 66.5 | 68.0 | 67.5 | 67.0 | 66.6 | 66.6 | 68.5 | 68.5 |
| 30 | 68.2 | 68.4 | 68.1 | 68.7 | 67.6 | 68.5 | 69.5 | 67.6 | 67.3 | 66.5 | 67.6 | 67.8 |
| 31 | 68.2 | 69.6 | 67.5 | 66.4 | 66.5 | 68.0 | 67.5 | 67.5 | 67.1 | 67.2 | 68.4 | 69.0 |

**January, 1889.**

for any hour added to the base-line value gives the absolute westerly declination at that hour.

= 2° 50′ 59′

| Day. | P. M. | | | | | | | | | | | | Mean. |
|---|---|---|---|---|---|---|---|---|---|---|---|---|---|
| | 1. | 2. | 3. | 4. | 5. | 6. | 7. | 8. | 9. | 10. | 11. | 12. | |
| | ′ | ′ | ′ | ′ | ′ | ′ | ′ | ′ | ′ | ′ | ′ | ′ | ′ |
| 1 | ... | ... | ... | ... | ... | ... | ... | ... | ... | ... | ... | ... | ... |
| 2 | ... | ... | ... | ... | ... | ... | ... | ... | ... | ... | ... | ... | ... |
| 3 | ... | ... | ... | ... | ... | ... | ... | ... | ... | ... | ... | ... | ... |
| 4 | ... | ... | ... | ... | ... | ... | ... | ... | ... | ... | ... | ... | ... |
| 5 | ... | ... | ... | ... | ... | ... | ... | ... | ... | ... | ... | ... | ... |
| 6 | ... | ... | ... | ... | ... | ... | ... | ... | ... | ... | ... | ... | ... |
| 7 | ... | ... | ... | ... | ... | ... | ... | ... | ... | ... | ... | ... | ... |
| 8 | ... | ... | ... | ... | ... | ... | ... | ... | ... | ... | ... | ... | ... |
| 9 | ... | ... | ... | 72.0 | 71.3 | 70.8 | 70.5 | 70.2 | 70.0 | 67.7 | 70.5 | 70.8 | 70.76 |
| 10 | ... | ... | ... | 70.2 | 69.8 | 70.5 | 69.5 | 66.7 | 69.5 | 69.4 | 69.6 | 70.0 | 70.03 |
| 11 | 74.5 | 73.2 | 73.4 | 73.5 | 70.5 | 71.4 | 71.0 | 70.6 | 70.5 | 68.8 | 70.4 | 72.5 | 70.86 |
| 12 | 70.8 | 70.9 | 72.0 | 72.3 | 69.5 | 70.3 | 70.3 | 69.0 | 68.4 | 69.0 | 70.6 | 70.6 | 70.68 |
| 13 | 70.5 | 71.5 | 71.8 | 71.6 | 70.7 | 69.7 | 68.7 | 69.4 | 69.7 | 70.3 | 70.5 | 70.5 | 70.11 |
| 14 | 70.4 | 71.4 | 71.6 | 71.6 | 70.5 | 70.5 | 69.9 | 69.6 | 69.6 | 69.3 | 69.5 | 69.6 | 69.97 |
| 15 | 71.5 | 72.0 | 71.4 | 70.5 | 70.1 | 69.7 | 69.5 | 69.5 | 69.5 | 69.6 | 70.0 | 70.4 | 70.18 |
| 16 | 69.6 | 69.5 | 68.5 | 67.0 | 66.7 | 66.5 | 66.5 | 66.3 | 66.5 | 66.7 | 67.0 | 67.0 | 68.57 |
| 17 | 69.6 | 69.5 | 68.3 | 66.7 | 66.3 | 66.3 | 66.1 | 66.3 | 66.5 | 66.5 | 66.8 | 66.7 | 66.88 |
| 18 | 71.3 | 70.4 | 68.5 | 67.5 | 67.5 | 67.5 | 67.3 | 67.5 | 67.5 | 67.5 | 67.7 | 68.3 | 67.19 |
| 19 | 71.6 | 70.8 | 69.3 | 67.6 | 67.5 | 67.5 | 67.5 | 67.5 | 67.5 | 68.5 | 68.5 | 67.6 | 68.12 |
| 20 | 76.3 | 73.5 | 73.6 | 70.2 | 69.3 | 69.0 | 67.5 | 67.1 | 61.4 | 67.5 | 66.5 | 71.2 | 67.96 |
| 21 | 70.0 | 69.6 | ... | 67.2 | 68.1 | 68.4 | 67.6 | 67.3 | 68.0 | 66.5 | 68.4 | 69.2 | 68.10 |
| 22 | 71.6 | 72.5 | 71.6 | 69.5 | 69.3 | 68.5 | 68.6 | 68.0 | 68.2 | 67.5 | 68.4 | 68.4 | 68.75 |
| 23 | ... | 71.8 | 73.5 | 71.0 | 68.5 | 68.0 | 67.6 | 67.5 | 67.1 | 67.3 | 68.0 | 68.3 | 68.70 |
| 24 | 72.4 | 70.6 | 69.5 | 68.5 | 68.0 | 68.0 | 68.1 | 67.6 | 67.4 | 67.5 | 68.3 | 68.6 | 68.26 |
| 25 | 71.6 | 71.8 | 71.6 | 70.4 | 68.6 | 68.5 | 68.9 | 68.2 | 68.0 | 65.4 | 67.5 | 68.1 | 68.44 |
| 26 | 71.3 | 71.2 | 70.6 | 69.6 | 68.5 | 68.4 | 68.1 | 68.0 | 68.0 | 67.7 | 67.6 | 68.0 | 68.54 |
| 27 | 70.0 | 70.1 | 69.6 | 69.4 | 68.6 | 68.2 | 67.8 | 67.5 | 66.6 | 67.5 | 67.5 | 68.2 | 68.15 |
| 28 | 70.5 | 70.3 | 69.6 | 69.5 | 68.8 | 68.5 | 68.2 | 68.0 | 68.0 | 68.0 | 68.2 | 68.5 | 68.13 |
| 29 | 69.2 | 68.9 | 69.0 | 69.2 | 68.8 | 69.5 | 68.1 | 67.8 | 67.8 | 67.7 | 68.0 | 08.4 | 68.25 |
| 30 | 68.6 | 68.0 | 69.2 | 69.9 | 69.0 | 68.8 | 69.0 | 67.6 | 67.5 | 68.3 | 68.0 | 68.4 | 68.25 |
| 31 | 69.5 | 69.0 | 68.5 | 68.5 | 68.2 | 68.0 | 68.0 | 68.4 | 68.3 | 69.5 | 67.4 | 68.5 | 68.11 |

TABLE V—Continued—*Dec*

Ordinates, expressed in minutes of arc, taken from the daily declination traces. The ordinate

Base-line value

| Day. | A. M. | | | | | | | | | | | |
|---|---|---|---|---|---|---|---|---|---|---|---|---|
| | 1. | 2. | 3. | 4. | 5. | 6. | 7. | 8. | 9. | 10. | 11. | 12. |
| | ′ | ′ | ′ | ′ | ′ | ′ | ′ | ′ | ′ | ′ | ′ | ′ |
| 1 | ... | ... | ... | ... | ... | ... | ... | ... | ... | ... | ... | ... |
| 2 | ... | ... | ... | ... | ... | ... | ... | ... | ... | ... | ... | ... |
| 3 | 67.0 | 68.1 | 66.9 | 67.7 | 67.6 | 67.2 | 67.0 | 66.6 | 66.0 | 65.6 | 67.0 | 69.0 |
| 4 | 68.3 | 68.6 | 68.5 | 68.6 | 68.2 | 67.7 | 67.5 | 66.9 | 66.5 | 66.4 | 67.6 | 69.5 |
| 5 | 68.2 | 68.5 | 69.0 | 68.0 | 68.0 | 67.9 | 67.5 | 66.6 | 65.2 | 65.5 | 66.5 | 68.7 |
| 6 | 68.2 | 68.2 | 68.5 | 68.5 | 67.6 | 68.0 | 67.2 | 65.8 | 64.1 | 65.1 | 67.1 | 69.0 |
| 7 | 69.0 | 68.5 | 68.7 | 68.5 | 68.5 | 68.9 | 68.6 | 69.2 | 68.0 | 68.3 | 69.5 | 71.0 |
| 8 | 71.7 | 69.5 | 69.5 | 68.8 | 74.0 | 70.9 | 68.0 | 67.1 | 65.6 | 66.5 | 68.0 | 69.0 |
| 9 | 68.5 | 68.9 | 68.7 | 68.6 | 68.5 | 68.2 | 68.0 | 67.4 | 66.2 | 67.3 | 67.5 | 70.0 |
| 10 | 69.1 | 69.1 | 69.0 | 69.0 | 68.6 | 68.3 | 68.0 | 68.5 | 67.1 | 67.5 | 68.4 | 69.5 |
| 11 | 68.7 | 68.8 | 68.7 | 68.6 | 68.5 | 68.5 | 68.4 | 68.1 | 68.0 | 68.4 | 69.8 | 70.7 |
| 12 | 68.9 | 68.9 | 69.5 | 68.9 | 68.5 | 68.1 | 68.0 | 67.0 | 67.0 | 68.3 | 70.8 | 72.6 |
| 13 | 69.0 | 68.9 | 69.0 | 68.9 | 68.6 | 68.5 | 68.1 | 68.0 | 68.5 | 68.2 | 69.8 | 71.0 |
| 14 | 68.7 | 68.6 | 68.4 | 67.3 | 67.5 | 64.7 | 65.0 | 66.0 | 67.0 | 68.5 | 71.1 | 73.2 |
| 15 | 69.3 | 68.7 | 68.4 | 65.6 | 68.4 | 66.6 | 68.0 | 69.4 | 69.3 | 68.1 | 71.0 | 71.7 |
| 16 | 67.9 | 69.0 | 67.9 | 69.4 | 67.8 | 68.0 | 67.4 | 66.9 | 67.2 | 69.4 | 72.2 | 72.3 |
| 17 | 68.0 | 68.2 | 69.8 | 67.0 | 66.9 | 66.6 | 66.3 | 73.3 | 72.0 | 66.5 | 69.4 | 70.5 |
| 18 | 65.3 | 68.4 | 68.8 | 68.7 | 67.4 | 69.5 | 68.7 | 67.6 | 66.7 | 67.4 | 69.3 | 69.5 |
| 19 | 69.2 | 68.8 | 68.7 | 69.2 | 69.0 | 69.3 | 68.6 | 67.6 | 67.0 | 66.5 | 67.0 | 69.0 |
| 20 | 67.6 | 70.5 | 70.0 | 72.2 | 66.7 | 68.0 | 67.3 | 66.8 | 67.1 | 67.6 | 68.6 | 71.5 |
| 21 | 68.7 | 68.9 | 69.2 | 69.0 | 69.0 | 69.0 | 68.6 | 67.3 | 67.3 | 67.5 | 68.3 | 69.9 |
| 22 | 68.5 | 68.6 | 68.6 | 68.6 | 68.3 | 68.0 | 69.0 | 66.7 | 67.0 | 67.3 | 67.8 | 69.4 |
| 23 | 68.1 | 68.8 | 68.8 | 68.9 | 69.0 | 68.8 | 69.1 | 67.4 | 67.3 | 67.5 | 68.4 | 69.0 |
| 24 | 68.5 | 69.0 | 69.4 | 69.0 | 69.4 | 68.9 | 68.8 | 67.2 | 67.2 | 67.0 | 68.1 | 69.4 |
| 25 | 69.3 | 69.4 | 69.6 | 69.4 | 69.1 | 68.8 | 69.0 | 68.4 | 67.8 | 68.0 | 67.4 | 68.4 |
| 26 | 68.8 | 68.9 | 68.8 | 68.6 | 68.6 | 68.4 | 68.3 | 67.6 | 67.4 | 67.1 | 67.5 | 68.8 |
| 27 | 68.6 | 66.6 | 68.6 | 68.6 | 67.9 | 67.9 | 67.5 | 67.6 | 67.2 | 68.3 | 67.5 | 68.7 |
| 28 | 68.9 | 68.9 | 68.6 | 68.5 | 68.2 | 68.1 | 67.6 | 67.0 | 68.3 | 68.2 | 68.9 | 70.2 |

*lination, February,* 1889.

for any hour added to the base-line value gives the absolute westerly declination at that hour.

$= 2° 51' 08''$

| Day. | P. M. 1. | 2. | 3. | 4. | 5. | 6. | 7. | 8. | 9. | 10. | 11. | 12. | Mean. |
|---|---|---|---|---|---|---|---|---|---|---|---|---|---|
| 1 | ... | ... | ... | ... | ... | ... | ... | ... | ... | ... | ... | ... | ... |
| 2 | 70.1 | 70.7 | 70.5 | 69.5 | 68.6 | 68.2 | 68.2 | 67.7 | 67.6 | 67.5 | 68.1 | 67.5 | ... |
| 3 | 70.6 | 72.3 | 71.5 | 71.2 | 68.5 | 68.0 | 68.0 | 67.6 | 67.5 | 67.7 | 68.0 | 68.1 | 68.11 |
| 4 | 70.7 | 71.6 | 71.4 | 70.0 | 69.1 | 67.9 | 67.2 | 67.7 | 67.5 | 67.9 | 68.0 | 68.0 | 68.39 |
| 5 | 70.1 | 70.1 | 70.0 | 69.6 | 69.5 | 69.1 | 68.7 | 68.1 | 68.0 | 67.6 | 67.8 | 67.0 | 68.08 |
| 6 | 70.5 | 70.8 | 70.7 | 70.8 | 69.1 | 68.7 | 69.1 | 69.9 | 69.0 | 70.0 | 70.1 | 69.0 | 68.54 |
| 7 | 69.8 | 69.5 | ... | ... | 70.4 | 70.6 | 69.8 | 67.1 | 67.7 | 67.4 | 67.1 | 68.0 | 68.82 |
| 8 | 71.6 | 71.0 | 69.8 | 69.5 | 68.2 | 68.9 | 68.5 | 68.3 | 68.5 | 68.4 | 68.5 | 68.5 | 69.10 |
| 9 | 71.1 | 71.0 | 70.4 | 70.1 | 69.6 | 68.9 | 68.6 | 68.6 | 68.5 | 67.4 | 68.6 | 69.0 | 68.73 |
| 10 | 70.0 | 70.1 | 70.0 | 69.5 | 69.0 | 68.5 | 68.5 | 68.1 | 68.1 | 68.2 | 68.4 | 68.2 | 68.70 |
| 11 | 71.5 | 71.6 | 70.8 | 69.3 | 68.8 | 68.8 | 68.7 | 68.5 | 68.4 | 68.2 | 68.0 | 68.6 | 69.05 |
| 12 | 72.4 | 71.6 | 70.1 | 69.5 | 69.0 | 68.5 | 68.1 | 68.0 | 68.0 | 68.0 | 68.2 | 68.6 | 69.02 |
| 13 | 71.6 | 70.6 | 69.9 | 68.0 | 68.9 | 69.0 | 68.7 | 68.6 | 68.4 | 68.5 | 69.0 | 69.0 | 69.03 |
| 14 | 72.0 | 71.3 | 70.3 | 70.6 | 71.4 | 72.0 | 73.0 | 68.0 | 68.7 | 69.1 | 69.1 | 69.5 | 69.21 |
| 15 | 72.0 | 72.9 | ... | ... | 69.1 | 69.0 | 68.8 | 68.5 | 66.4 | 68.0 | 66.7 | 67.4 | 68.79 |
| 16 | 71.8 | 71.7 | 71.0 | 71.1 | 70.6 | 69.5 | 68.9 | 68.4 | 68.3 | 68.4 | 68.3 | 68.2 | 69.24 |
| 17 | 71.1 | 71.0 | 73.4 | 70.4 | 70.5 | 69.0 | 69.6 | 68.6 | 65.3 | 66.6 | 68.7 | 64.9 | 68.90 |
| 18 | 71.0 | 71.6 | 71.5 | 69.4 | 69.2 | 68.4 | 69.4 | 68.4 | 65.5 | 68.7 | 68.8 | 68.3 | 68.65 |
| 19 | 70.8 | 71.6 | 72.2 | 72.8 | 70.2 | 70.0 | 68.6 | 68.4 | 67.9 | 69.0 | 66.9 | 65.9 | 68.93 |
| 20 | 71.9 | 72.4 | 71.0 | 71.0 | 70.3 | 69.6 | 70.0 | 68.8 | 68.8 | 68.8 | 69.1 | 69.0 | 69.36 |
| 21 | 71.1 | 71.8 | 71.4 | 71.3 | 70.6 | 68.6 | 69.2 | 67.4 | 68.8 | 68.5 | 68.8 | 68.5 | 69.11 |
| 22 | 70.8 | 71.1 | 71.1 | 70.5 | 70.0 | 70.7 | 69.9 | 70.1 | 66.4 | 68.1 | 68.8 | 65.5 | 68.78 |
| 23 | 69.9 | 71.1 | ... | ... | 70.5 | 70.4 | 69.2 | 69.0 | 68.7 | 68.1 | 68.0 | 68.9 | 68.86 |
| 24 | 70.5 | 71.4 | 71.3 | 70.6 | 68.9 | 69.4 | 69.1 | 69.0 | 69.2 | 68.9 | 68.9 | 69.0 | 69.09 |
| 25 | 69.5 | 70.2 | 70.4 | ... | 69.4 | 68.5 | 68.9 | 68.7 | 68.6 | 68.6 | 68.9 | 68.9 | 68.55 |
| 26 | 70.5 | 71.2 | 71.4 | 71.9 | 71.9 | 71.1 | 70.3 | 69.0 | 68.3 | 64.1 | 68.6 | 68.6 | 68.99 |
| 27 | 71.7 | 70.6 | 70.7 | 69.9 | 68.5 | 69.1 | 68.6 | 64.7 | 67.6 | 67.7 | 69.4 | 69.9 | 68.47 |
| 28 | 71.9 | 71.4 | 71.0 | 70.0 | 70.2 | 71.0 | 69.6 | 69.7 | 68.6 | 67.6 | 67.6 | 59.8 | 69.13 |

TABLE V—Continued—*Dec*

Ordinates, expressed in minutes of arc, taken from the daily declination traces.　The ordinate

Base-line value

| Day. | A. M. | | | | | | | | | | | |
|---|---|---|---|---|---|---|---|---|---|---|---|---|
| | 1. | 2. | 3. | 4. | 5. | 6. | 7. | 8. | 9. | 10. | 11. | 12. |
| 1 | 64.2 | 68.1 | 68.4 | 67.6 | 68.2 | 69.1 | 71.6 | 77.4 | 71.4 | 72.0 | 69.0 | 70.5 |
| 2 | 68.2 | 67.4 | 67.7 | 67.3 | 67.6 | 69.6 | 69.0 | 67.0 | 66.3 | 68.1 | 69.9 | 72.1 |
| 3 | 69.0 | 69.9 | 70.4 | 68.4 | 67.5 | 67.8 | 67.4 | 67.0 | 66.4 | 67.0 | 68.9 | 70.4 |
| 4 | 68.7 | 67.6 | 67.1 | 67.1 | 67.0 | 66.9 | 68.0 | 65.7 | 66.0 | 67.8 | 68.9 | 70.7 |
| 5 | 67.7 | 69.0 | 67.1 | 67.9 | 69.1 | 67.4 | 67.3 | 66.8 | 67.5 | 67.8 | 68.4 | 69.8 |
| 6 | 70.2 | 66.3 | 64.2 | 67.1 | 73.5 | 72.0 | 69.1 | 71.0 | 71.6 | 75.5 | 74.6 | . . . |
| 7 | 66.6 | 66.8 | 70.1 | 68.4 | 69.2 | 69.4 | 68.0 | 66.4 | 66.6 | 67.8 | 67.7 | 69.6 |
| 8 | 68.4 | 71.1 | 71.1 | 68.1 | 68.1 | 68.4 | 68.1 | 67.0 | 68.0 | 67.9 | 69.9 | 72.3 |
| 9 | 68.9 | 69.0 | 69.8 | 69.0 | 69.0 | 68.6 | 68.0 | 67.0 | 66.7 | 67.7 | 69.4 | . . . |
| 10 | 69.2 | 68.9 | 68.7 | 68.6 | 68.2 | 68.9 | 68.4 | 66.4 | 66.0 | 66.2 | 68.0 | 70.2 |
| 11 | 69.0 | 69.2 | 68.2 | 68.2 | 68.4 | 68.4 | 68.1 | 67.1 | 67.4 | 67.8 | 69.1 | . . . |
| 12 | 68.6 | 66.9 | 67.0 | 67.0 | 66.5 | 67.9 | 67.5 | 66.9 | 65.4 | 67.0 | 68.2 | 70.0 |
| 13 | 68.1 | 69.1 | 68.4 | 68.2 | 68.2 | 68.1 | 67.6 | 67.3 | 66.2 | 66.0 | 67.7 | 69.3 |
| 14 | 69.0 | 68.3 | 68.3 | 68.8 | 68.6 | 68.6 | 68.0 | 67.3 | 67.2 | 67.7 | 69.2 | 71.4 |
| 15 | 69.1 | 69.1 | 68.8 | 68.3 | 67.8 | 68.0 | 66.1 | 66.2 | 65.2 | 66.6 | 69.5 | 71.0 |
| 16 | 67.7 | 72.1 | 68.3 | 67.7 | 67.8 | 68.0 | 67.0 | 67.0 | 66.7 | 67.7 | 70.2 | 71.6 |
| 17 | 68.2 | 68.1 | 68.0 | 68.0 | 67.5 | 67.9 | 66.1 | 64.3 | 64.4 | 66.2 | 69.8 | 71.7 |
| 18 | 67.1 | 68.2 | 68.0 | 68.5 | 69.0 | 68.0 | 65.9 | 64.7 | 65.2 | . . . | 69.9 | . . . |
| 19 | 69.0 | 68.9 | 68.3 | 68.1 | 68.0 | 67.3 | 66.3 | 65.0 | 64.9 | 65.9 | 67.8 | 69.9 |
| 20 | 68.4 | 68.8 | 68.1 | 68.0 | 67.6 | 68.0 | 66.3 | 65.7 | 65.6 | 65.5 | 68.0 | 69.9 |
| 21 | 69.8 | 68.8 | 68.3 | 68.9 | 68.6 | 68.4 | 67.1 | 66.6 | 67.1 | 66.8 | 68.7 | 72.0 |
| 22 | 66.3 | 64.0 | 67.2 | 64.3 | 65.6 | 65.9 | 65.5 | 65.2 | 66.0 | 66.1 | 69.5 | 71.6 |
| 23 | 68.7 | 68.6 | 70.0 | 68.2 | 68.1 | 68.0 | 68.0 | 67.4 | 66.6 | 66.5 | 68.4 | 69.7 |
| 24 | 68.8 | 68.2 | 68.1 | 68.0 | 67.2 | 69.1 | 70.9 | 67.6 | 67.0 | 67.6 | 69.1 | 70.8 |
| 25 | 66.2 | 68.3 | 67.5 | 67.9 | 67.0 | 68.1 | 67.0 | 66.7 | 67.1 | 67.2 | 68.8 | 70.7 |
| 26 | 69.1 | 68.5 | 67.9 | 67.9 | 67.6 | 67.6 | 67.1 | 66.2 | 65.8 | 66.2 | 68.8 | 72.1 |
| 27 | 67.6 | 68.0 | 68.1 | 68.2 | 68.1 | 67.0 | 66.1 | 65.4 | 66.0 | 66.4 | 69.2 | 71.0 |
| 28 | 60.3 | 68.8 | 64.4 | 67.4 | 68.8 | 66.7 | 66.0 | 69.4 | 64.0 | 66.0 | 70.1 | 75.0 |
| 29 | 68.4 | 69.2 | 68.7 | 68.7 | 68.8 | 67.6 | 65.1 | 64.5 | 66.4 | 66.4 | 69.9 | 74.4 |
| 30 | 68.4 | 70.2 | 67.2 | 66.1 | 68.2 | 68.8 | 66.7 | 64.5 | 64.9 | 65.8 | 67.4 | 68.8 |
| 31 | 69.2 | 68.9 | 68.8 | 69.5 | 70.1 | 68.2 | 66.9 | 65.0 | 65.1 | 65.9 | 68.2 | 71.4 |

*lination, March,* 1889

for any hour added to the base-line value gives the absolute westerly declination at that hour.

$= 2° 51' 31''$

| Day. | P. M. | | | | | | | | | | | | Mean. |
|---|---|---|---|---|---|---|---|---|---|---|---|---|---|
| | 1. | 2. | 3. | 4. | 5. | 6. | 7. | 8. | 9. | 10. | 11. | 12. | |
| 1 | 74.0 | 75.7 | 73.2 | 70.3 | 69.0 | 68.9 | 68.7 | 68.5 | 68.7 | 69.9 | 69.1 | 69.3 | 70.01 |
| 2 | 73.2 | 74.0 | 73.6 | 71.5 | 69.9 | 69.3 | 68.9 | 68.8 | 68.9 | 68.9 | 69.0 | 69.2 | 69.39 |
| 3 | 72.3 | 72.5 | 72.4 | 71.3 | 70.3 | 69.6 | 69.3 | 69.1 | 69.3 | 69.0 | 69.6 | 70.3 | 69.38 |
| 4 | 71.7 | 71.8 | 71.4 | 70.3 | 69.6 | 69.5 | 69.1 | 68.9 | 68.9 | 69.0 | 68.7 | 68.6 | 68.71 |
| 5 | 70.7 | 70.9 | 70.5 | 69.8 | 69.9 | 70.3 | 69.9 | 69.2 | 69.5 | 64.0 | 66.1 | 66.0 | 68.44 |
| 6 | 72.6 | 72.6 | 71.1 | 70.6 | 69.9 | 69.1 | 68.6 | 67.7 | 68.1 | 67.1 | 65.9 | 64.4 | 69.69 |
| 7 | 71.1 | 72.0 | 72.3 | 70.8 | 70.1 | 68.9 | 69.0 | 68.9 | 68.6 | 67.2 | 65.7 | 68.7 | 68.75 |
| 8 | 73.1 | 72.5 | 72.1 | 70.7 | 70.2 | 69.5 | 69.0 | 68.6 | 67.6 | 66.2 | 68.7 | 69.0 | 69.40 |
| 9 | 72.4 | 72.7 | 71.9 | 70.8 | 69.7 | 69.4 | 69.2 | 69.0 | 69.0 | 69.0 | 69.0 | 68.6 | 69.32 |
| 10 | 71.8 | 72.1 | 71.9 | 71.0 | 69.9 | 69.2 | 69.1 | 69.0 | 69.3 | 69.0 | 69.0 | 68.6 | 69.07 |
| 11 | 70.5 | 71.1 | 71.3 | 71.3 | 72.1 | 71.7 | 70.0 | 69.2 | 69.1 | 68.7 | 68.4 | 68.5 | 69.14 |
| 12 | 70.9 | 71.0 | 70.3 | 69.5 | 69.5 | 69.6 | 69.3 | 69.0 | 69.0 | 68.3 | 68.7 | 68.5 | 68.44 |
| 13 | 71.2 | 71.9 | 73.1 | 77.0 | 72.9 | 73.2 | 73.3 | 73.0 | 70.0 | 63.5 | 68.5 | 68.9 | 69.61 |
| 14 | 73.2 | 72.1 | 71.9 | 71.1 | 70.1 | 69.9 | 69.0 | 69.0 | 69.0 | 68.5 | 67.4 | 68.4 | 69.25 |
| 15 | 72.3 | 73.3 | 73.0 | 70.7 | 69.7 | 69.4 | 69.0 | 68.7 | 68.8 | 69.0 | 69.7 | 67.8 | 69.05 |
| 16 | 72.2 | 72.6 | 72.3 | 70.9 | 69.6 | 68.6 | 68.3 | 68.6 | 68.5 | 68.7 | 68.6 | 68.5 | 69.13 |
| 17 | 74.1 | 73.1 | 74.1 | 73.5 | 73.9 | 71.0 | 72.0 | 71.8 | 68.0 | 66.9 | 71.2 | 63.8 | 69.32 |
| 18 | 72.0 | 72.1 | 72.0 | 71.0 | 70.0 | 69.3 | 69.0 | 68.9 | 68.7 | 67.0 | 68.7 | 67.3 | 68.66 |
| 19 | 71.8 | 72.5 | 72.1 | 71.6 | 70.6 | 69.6 | 69.2 | 68.7 | 68.8 | 64.9 | 67.3 | 69.0 | 68.56 |
| 20 | 71.6 | 72.7 | 73.5 | 73.1 | 73.4 | 70.8 | 70.5 | 69.2 | 69.0 | 67.8 | 69.1 | 69.0 | 69.15 |
| 21 | 73.9 | 74.4 | 75.2 | 74.0 | 72.0 | 70.6 | 69.9 | 68.8 | 69.5 | 69.1 | 68.8 | 68.9 | 69.84 |
| 22 | 72.9 | 72.1 | 73.4 | 71.8 | 71.9 | 70.9 | 63.5 | 68.9 | 70.2 | 69.4 | 68.9 | 68.8 | 68.33 |
| 23 | 70.8 | 71.0 | 70.7 | 70.4 | 69.7 | 69.4 | 69.2 | 68.9 | 69.0 | 68.8 | 69.3 | 68.7 | 68.92 |
| 24 | 71.1 | 71.5 | 71.4 | 71.0 | 70.4 | 70.4 | 69.8 | 69.6 | 69.8 | 68.3 | 69.9 | 71.0 | 69.44 |
| 25 | 71.0 | 71.3 | 71.7 | 70.4 | 70.0 | 69.7 | 69.7 | 69.1 | 69.2 | 69.1 | 69.0 | 68.9 | 68.82 |
| 26 | 71.8 | 70.8 | 71.4 | 70.8 | 70.5 | 70.2 | 68.9 | 66.8 | 68.6 | 68.3 | . . . | . . . | 68.77 |
| 27 | 71.6 | 72.5 | 72.4 | 71.4 | 70.5 | 70.6 | 70.1 | 70.7 | 64.9 | 66.3 | 57.4 | 65.6 | 68.55 |
| 28 | 73.6 | 75.9 | 72.2 | 73.2 | 71.2 | 69.7 | 69.9 | 66.3 | 65.2 | 67.5 | 68.4 | 68.4 | 68.68 |
| 29 | 74.3 | 73.6 | 73.9 | 72.6 | 70.4 | 68.5 | 69.6 | 69.4 | 69.2 | 69.3 | 67.2 | 68.2 | 69.35 |
| 30 | 70.5 | 71.8 | 72.4 | 71.8 | 70.6 | 69.5 | 69.1 | 68.6 | 69.0 | 68.9 | 68.9 | 69.0 | 68.63 |
| 31 | 73.4 | 74.5 | 74.6 | 73.0 | 71.1 | 70.3 | 69.6 | 69.3 | 68.9 | 69.0 | 69.0 | 68.9 | 69.53 |

TABLE V—Continued—*Dec*

Ordinates, expressed in minutes of arc, taken from the daily declination traces. The ordinate

Base-line value

| Day. | A. M. | | | | | | | | | | | |
|---|---|---|---|---|---|---|---|---|---|---|---|---|
| | 1. | 2. | 3. | 4. | 5. | 6. | 7. | 8. | 9. | 10. | 11. | 12. |
| 1 | 69.3 | 68.9 | 68.8 | 69.6 | 70.1 | 68.4 | 66.9 | 65.1 | 65.1 | 65.9 | 63.3 | 71.4 |
| 2 | 64.9 | 68.1 | 68.0 | 67.4 | 67.2 | 66.4 | 66.7 | 65.6 | 65.7 | 67.6 | 69.3 | 72.6 |
| 3 | 67.6 | 65.9 | 65.0 | 69.1 | 67.0 | 65.7 | 65.4 | 65.1 | 64.9 | 64.2 | 67.0 | 69.6 |
| 4 | 69.5 | 69.0 | 70.0 | 69.2 | 69.1 | 68.5 | 66.9 | 65.2 | 66.0 | 67.9 | 71.2 | 74.0 |
| 5 | 71.0 | 71.5 | 70.9 | 70.2 | 70.7 | 70.2 | 69.2 | 68.0 | 68.0 | 68.0 | ... | 69.0 |
| 6 | 68.8 | 68.8 | 68.8 | 69.0 | 69.0 | 68.6 | 68.1 | 67.0 | 66.5 | 67.1 | 68.1 | 69.6 |
| 7 | 68.5 | 68.0 | 67.5 | 67.1 | 67.1 | 66.5 | 67.0 | 67.0 | 66.0 | 68.0 | 68.5 | 70.9 |
| 8 | 69.6 | 70.0 | 57.7 | 72.8 | 67.5 | 66.8 | 66.6 | 67.1 | 67.6 | 70.0 | 72.2 | 71.5 |
| 9 | 69.1 | 70.6 | 69.0 | 69.1 | 68.6 | 68.0 | 67.9 | 66.0 | 67.5 | 69.0 | 71.5 | 73.5 |
| 10 | 69.6 | 69.2 | 68.6 | 71.9 | 68.8 | 67.0 | 67.5 | 66.9 | 68.4 | 68.0 | 68.9 | 71.0 |
| 11 | 68.7 | 68.0 | 68.0 | 67.8 | 67.8 | 68.1 | 67.0 | 67.0 | 67.1 | 68.0 | 69.0 | 69.9 |
| 12 | 69.5 | 69.3 | 69.0 | 69.0 | 68.6 | 68.0 | 67.4 | 67.4 | 68.1 | 69.9 | 71.0 | 72.2 |
| 13 | 67.9 | 68.8 | 68.5 | 68.5 | 68.6 | 67.9 | 67.3 | 67.0 | 67.8 | 69.9 | 71.4 | 72.6 |
| 14 | 68.7 | 69.0 | 68.9 | 68.8 | 68.1 | 67.5 | 66.9 | 66.8 | 67.1 | 68.0 | 71.6 | 74.0 |
| 15 | 69.2 | 69.0 | 68.5 | 68.5 | 68.4 | 67.6 | 67.1 | 66.1 | 66.5 | 69.0 | 72.5 | 74.3 |
| 16 | 69.2 | 69.1 | 68.5 | 68.0 | 67.6 | 66.4 | 64.6 | 63.4 | 64.5 | 68.4 | 71.6 | 73.1 |
| 17 | 69.4 | 69.0 | 68.3 | 68.1 | 67.5 | 66.5 | 65.0 | 63.8 | 64.2 | 69.0 | 71.2 | 72.5 |
| 18 | 69.2 | 68.9 | 68.4 | 67.7 | 67.9 | 66.1 | 65.7 | 65.1 | 66.0 | 69.2 | 72.0 | 72.6 |
| 19 | 68.2 | 68.0 | 68.0 | 67.8 | 68.6 | 67.5 | 66.2 | 66.5 | 67.0 | 68.9 | 70.1 | 70.8 |
| 20 | 70.1 | 70.0 | 69.2 | 69.2 | 68.6 | 69.0 | 68.5 | 68.1 | 68.0 | 69.0 | 71.1 | 72.5 |
| 21 | 69.8 | 68.6 | 69.0 | 69.5 | 69.1 | 68.7 | 68.1 | 67.5 | 66.7 | 68.5 | 71.6 | 74.7 |
| 22 | 69.9 | 68.5 | 69.7 | 69.4 | 68.7 | 67.2 | 66.4 | 66.8 | 67.9 | 69.1 | 71.2 | 73.1 |
| 23 | 69.9 | 69.5 | 69.5 | 68.6 | 69.4 | 67.8 | 67.6 | 67.4 | 69.1 | 71.6 | 73.0 | 75.1 |
| 24 | 70.1 | 70.0 | 69.8 | 69.5 | 68.5 | 68.0 | 67.9 | 67.0 | 68.0 | 69.8 | 71.2 | 72.3 |
| 25 | 69.3 | 69.1 | 69.0 | 68.9 | 68.8 | 67.3 | 65.7 | 71.7 | 69.5 | 72.0 | 74.0 | 74.5 |
| 26 | 71.2 | 70.5 | 69.8 | 69.9 | 69.0 | 67.5 | 67.0 | 66.9 | 68.7 | 70.0 | 73.0 | 74.6 |
| 27 | 70.5 | 71.0 | 71.7 | 69.4 | 69.6 | 67.0 | 65.1 | 64.5 | 65.6 | 68.0 | 71.2 | 74.4 |
| 28 | 68.2 | 69.1 | 68.5 | 67.2 | 67.2 | 64.0 | 63.6 | 61.7 | 64.0 | 68.1 | 71.9 | 74.9 |
| 29 | 69.0 | 69.2 | 69.5 | 69.6 | 68.9 | 67.1 | 66.1 | 65.9 | 66.0 | 69.2 | 71.0 | 72.5 |
| 30 | 69.6 | 69.0 | 70.7 | 68.9 | 67.6 | 67.4 | 68.5 | 68.2 | 68.5 | 69.3 | 71.4 | 72.9 |

*lination; April,* 1889.

for any hour added to the base-line value gives the absolute westerly declination at that hour.

= 2° 51′ 33″

| Day. | P. M. | | | | | | | | | | | | Mean. |
|---|---|---|---|---|---|---|---|---|---|---|---|---|---|
| | 1. | 2. | 3. | 4. | 5. | 6. | 7. | 8. | 9. | 10. | 11. | 12. | |
| 1 | 72.0 | 73.8 | 75.1 | 76.0 | 72.1 | 70.2 | 69.2 | 68.7 | 68.2 | 68.9 | 68.4 | 63.9 | 69.35 |
| 2 | 74.0 | 74.4 | 73.3 | 71.7 | 69.8 | 69.0 | 68.9 | 68.8 | 68.2 | 68.2 | 65.4 | 68.5 | 68.74 |
| 3 | 73.0 | 74.2 | 74.1 | 73.5 | 70.5 | 69.5 | 69.5 | 69.5 | 69.0 | 68.4 | 69.5 | 69.8 | 68.63 |
| 4 | 75.0 | 74.0 | 73.0 | ... | ... | 71.2 | 71.5 | 71.2 | 71.2 | 71.2 | 71.3 | 71.0 | 70.32 |
| 5 | 70.0 | 71.6 | 72.6 | 72.5 | 17.6 | 70.0 | 69.0 | 68.9 | 69.1 | 69.1 | 69.1 | 68.5 | 69.94 |
| 6 | 70.8 | 71.4 | 71.5 | 71.3 | 70.2 | 70.0 | 69.3 | 68.9 | 68.8 | 69.1 | 68.9 | 69.0 | 69.11 |
| 7 | 72.0 | 73.5 | 75.0 | 74.1 | 72.6 | 70.9 | 70.5 | 70.1 | 65.2 | 55.8 | 68.0* | 67.1 | 69.18 |
| 8 | 71.2 | 72.2 | 73.0 | 71.8 | 71.1 | 71.1 | 70.0 | 65.1 | 69.2 | 69.3 | 68.6 | 69.3 | 69.78 |
| 9 | 74.5 | 77.0 | 72.9 | 72.9 | 71.5 | 70.1 | 66.0 | 69.2 | 67.8 | 68.5 | 68.1 | 68.8 | 69.88 |
| 10 | 71.0 | 71.1 | 71.5 | 71.5 | 70.8 | 70.0 | 69.5 | 69.4 | 69.5 | 69.1 | 68.5 | 68.7 | 69.43 |
| 11 | 70.1 | 70.5 | 71.5 | 71.0 | 70.1 | 69.6 | 69.2 | 69.5 | 69.1 | 69.5 | 69.5 | 69.5 | 68.98 |
| 12 | 72.7 | 72.5 | 71.6 | 71.3 | 71.2 | 70.9 | 69.5 | 70.0 | 69.5 | 69.0 | 69.5 | 68.0 | 69.80 |
| 13 | 73.0 | 72.6 | 72.5 | 71.3 | 71.0 | 70.2 | 70.0 | 69.9 | 69.4 | 69.5 | 69.5 | 69.0 | 69.78 |
| 14 | 73.8 | 73.3 | 73.1 | 72.0 | 70.8 | 70.0 | 69.0 | 69.8 | 66.8 | 69.9 | 69.8 | 69.5 | 69.72 |
| 15 | 75.1 | 75.8 | 75.4 | 73.7 | 71.4 | 69.6 | 69.8 | 68.1 | 68.3 | 69.6 | 69.5 | 69.5 | 70.01 |
| 16 | 74.5 | 75.0 | 73.6 | 72.4 | 70.8 | 69.5 | 69.5 | 69.5 | 69.6 | 69.8 | 69.1 | 69.1 | 69.45 |
| 17 | 72.6 | 72.5 | 72.0 | 71.0 | 70.0 | 69.5 | 69.7 | 69.5 | 68.4 | 68.9 | 68.6 | 68.1 | 68.97 |
| 18 | 73.3 | 73.5 | 72.9 | 72.0 | 70.9 | 69.9 | 69.5 | 69.5 | 70.0 | 69.2 | 69.0 | 68.4 | 69.45 |
| 19 | 73.0 | 74.1 | 74.2 | 74.0 | 73.0 | 71.8 | 70.8 | 70.6 | 70.6 | 66.5 | 69.9 | 69.5 | 69.82 |
| 20 | 73.5 | 74.0 | 73.6 | 72.5 | 71.6 | 70.2 | 70.4 | 70.0 | 69.9 | 69.7 | 70.0 | 70.0 | 70.04 |
| 21 | 75.4 | 75.3 | 74.9 | 74.0 | 72.4 | 71.4 | 71.3 | 71.0 | 70.9 | 69.3 | 69.1 | 67.9 | 70.06 |
| 22 | 74.4 | 74.3 | 74.2 | 73.5 | 72.2 | 71.5 | 70.9 | 71.0 | 68.7 | 68.5 | 69.5 | 69.2 | 70.02 |
| 23 | 73.1 | 73.0 | 73.1 | 73.2 | 73.1 | 72.0 | 71.5 | 68.6 | 69.0 | 66.9 | 67.0 | 70.0 | 70.04 |
| 24 | 72.9 | 72.7 | 72.3 | 72.0 | 71.8 | 71.0 | 70.7 | 70.6 | 70.6 | 70.5 | 70.1 | 69.0 | 70.03 |
| 25 | 75.0 | 75.8 | 75.0 | 75.0 | 74.2 | 71.0 | 72.2 | 70.0 | 67.8 | 71.1 | 70.6 | 72.0 | 71.23 |
| 26 | 75.1 | 74.6 | 73.5 | 73.0 | 71.8 | 71.6 | 71.0 | 69.9 | 70.5 | 70.8 | 71.0 | 70.9 | 71.09 |
| 27 | 76.4 | 76.5 | 74.7 | 72.5 | 71.3 | 69.8 | 69.0 | 70.5 | 67.0 | 58.8 | 60.2 | 63.9 | 69.56 |
| 28 | 75.1 | 75.0 | 74.2 | 73.5 | 71.0 | 70.1 | 70.0 | 68.2 | 70.1 | 69.1 | 68.5 | 69.8 | 69.29 |
| 29 | 73.2 | 73.8 | 73.6 | 73.4 | 72.1 | 71.5 | 71.0 | 70.0 | 68.5 | 70.0 | 68.6 | 69.9 | 69.98 |
| 30 | 73.0 | 73.4 | 72.9 | 72.5 | 71.5 | 70.0 | 70.0 | 66.6 | 68.8 | 68.9 | 70.1 | 70.1 | 69.99 |

TABLE V—Continued—*Dec*

Ordinates, expressed in minutes of arc, taken from the daily declination traces. The ordinate

Base-line value

| Day. | A. M. | | | | | | | | | | | |
|---|---|---|---|---|---|---|---|---|---|---|---|---|
| | 1. | 2. | 3. | 4. | 5. | 6. | 7. | 8. | 9. | 10. | 11. | 12. |
| 1 | 69.5 | 69.2 | 68.6 | 68.6 | 68.1 | 66.6 | 65.0 | 66.1 | 66.7 | 69.1 | 71.5 | 74.1 |
| 2 | 70.1 | 70.0 | 69.7 | 69.5 | 68.8 | 67.7 | 66.5 | 66.1 | 67.0 | 69.3 | 72.0 | 74.0 |
| 3 | 70.5 | 70.1 | 70.0 | 69.5 | 69.0 | 67.6 | 66.1 | 65.5 | 66.7 | 70.0 | 73.1 | 75.6 |
| 4 | 74.0 | 66.4 | 68.1 | 68.1 | 68.2 | 66.9 | 68.1 | 66.1 | 68.3 | 70.2 | 72.9 | 75.5 |
| 5 | 65.5 | 65.3 | 65.6 | 66.6 | 66.2 | 64.6 | 67.0 | 65.1 | 66.7 | 69.1 | 72.6 | 74.2 |
| 6 | 68.5 | 69.2 | 69.4 | 63.9 | 69.1 | 67.0 | 66.2 | 66.9 | 67.9 | 70.6 | 72.4 | 73.1 |
| 7 | 66.8 | 68.1 | 71.4 | 67.1 | 66.1 | 65.3 | 65.3 | 66.9 | 69.0 | 72.4 | 74.6 | 74.9 |
| 8 | 69.9 | 69.5 | 69.2 | 69.0 | 67.6 | 67.0 | 66.6 | 67.5 | 69.4 | 72.1 | 74.2 | 75.0 |
| 9 | 70.5 | 70.1 | 70.0 | 69.5 | 69.4 | 69.0 | 68.0 | 68.6 | 70.0 | 72.9 | 74.9 | 76.2 |
| 10 | 70.5 | 69.9 | 70.0 | 68.6 | 68.4 | 68.0 | 68.6 | 67.5 | 67.0 | 72.0 | 73.0 | 73.8 |
| 11 | 70.5 | 69.5 | 69.5 | 69.5 | 69.5 | 69.5 | 67.8 | 67.5 | 68.9 | 71.2 | 73.6 | . . . |
| 12 | 70.2 | 70.0 | 70.0 | 69.6 | 69.1 | 68.2 | 67.9 | 67.9 | 69.0 | 71.9 | 75.0 | 76.2 |
| 13 | 67.5 | 71.1 | 69.5 | 67.0 | 68.2 | 67.7 | 67.8 | 67.8 | 68.3 | 72.2 | 73.6 | 74.4 |
| 14 | 73.1 | 75.0 | 73.0 | 72.3 | 71.9 | 71.3 | 70.8 | 70.5 | 71.1 | 73.3 | 77.7 | 80.0 |
| 15 | 73.9 | 73.6 | 73.5 | 73.1 | 72.3 | 70.8 | 70.1 | 69.3 | 71.1 | 74.3 | 78.0 | 80.1 |
| 16 | 70.7 | 70.5 | 70.3 | 69.9 | 69.1 | 67.7 | 66.2 | 66.1 | 67.0 | 69.2 | 72.1 | 75.0 |
| 17 | 71.0 | 70.6 | 70.6 | 70.2 | 69.9 | 68.2 | 67.6 | 67.5 | 69.0 | 70.3 | 72.5 | 75.0 |
| 18 | 70.2 | 70.7 | 70.4 | 70.4 | 69.1 | 67.6 | 65.8 | 65.8 | 67.1 | 69.2 | 70.9 | 72.1 |
| 19 | 70.0 | 70.5 | 66.4 | 67.2 | 67.7 | 66.8 | 66.5 | 66.0 | 67.1 | 69.0 | 72.1 | 75.1 |
| 20 | 70.9 | 70.5 | 70.4 | 69.5 | 68.7 | 67.2 | 66.6 | 67.4 | 68.1 | 70.6 | 73.0 | 74.0 |
| 21 | 70.5 | 70.9 | 70.5 | 70.0 | 68.9 | 67.6 | 67.5 | 67.5 | 68.9 | 70.9 | 72.6 | 74.2 |
| 22 | 65.1 | 64.2 | . . . | . . . | . . . | . . : | . . . | . . . | . . . | . . . | . . . | . . . |
| 23 | 71.2 | 70.6 | 70.1 | 69.7 | 68.8 | 67.3 | 66.3 | 66.3 | 67.9 | 70.5 | 74.1 | 76.1 |
| 24 | 70.7 | 69.9 | 69.1 | 68.5 | 66.6 | 64.0 | 63.6 | 65.8 | 67.0 | 70.3 | 73.7 | 75.5 |
| 25 | 70.8 | 70.8 | 70.1 | 69.9 | 69.3 | 68.2 | 66.7 | 67.0 | 69.1 | 72.0 | 74.0 | 76.0 |
| 26 | 69.5 | 68.0 | 67.0 | 66.0 | 64.0 | 63.0 | 68.0 | 66.6 | 65.7 | 69.2 | 72.1 | 77.1 |
| 27 | 70.0 | 75.2 | 72.9 | 70.7 | 69.5 | 68.0 | 67.2 | 68.1 | 69.5 | 71.2 | 72.0 | 73.0 |
| 28 | 71.4 | 71.1 | 73.1 | 71.3 | 70.1 | 68.4 | 68.1 | 67.5 | 68.6 | 70.6 | 73.4 | 76.1 |
| 29 | 70.7 | 71.0 | 71.0 | 71.0 | 70.6 | 70.5 | 68.2 | 66.2 | 66.4 | 68.4 | 71.8 | 74.0 |
| 30 | 71.1 | 71.3 | 70.8 | 70.2 | 70.4 | 69.3 | 68.0 | 66.7 | 65.8 | 67.9 | 70.9 | 73.9 |
| 31 | 71.4 | 71.6 | 70.6 | 70.0 | 68.1 | 66.9 | 65.0 | 68.0 | 66.5 | 68.3 | 70.0 | 74.0 |

*lination, May,* 1889.

for any hour added to the base-line value gives the absolute westerly declination at that hour.

$=2°\ 50'\ 38''$

| Day. | P. M. | | | | | | | | | | | | Mean. |
|---|---|---|---|---|---|---|---|---|---|---|---|---|---|
| | 1. | 2. | 3. | 4. | 5. | 6. | 7. | 8. | 9. | 10. | 11. | 12. | |
| 1 | 74.5 | 74.1 | 73.5 | 72.5 | ... | ... | 70.5 | 70.1 | 70.1 | 67.8 | 68.2 | 69.6 | 69.77 |
| 2 | 74.9 | 74.9 | 73.9 | 72.4 | 71.1 | 69.3 | 70.2 | 70.2 | 70.2 | 70.2 | 70.5 | 70.5 | 70.37 |
| 3 | 76.0 | 75.3 | 74.0 | 72.0 | 70.6 | 70.2 | 70.4 | 71.1 | 71.1 | 70.7 | 70.5 | 68.1 | 70.57 |
| 4 | 75.1 | 75.7 | 76.1 | 74.9 | 72.6 | 71.1 | 70.5 | 70.0 | 70.0 | 69.8 | 65.5 | 68.0 | 70.50 |
| 5 | 75.0 | 75.0 | 74.0 | 73.5 | 72.0 | 70.6 | 70.9 | 70.2 | 70.4 | 69.9 | 69.2 | 66.1 | 69.39 |
| 6 | 73.1 | 72.6 | 71.5 | 70.1 | 69.9 | 69.6 | 70.0 | 71.0 | 70.0 | 70.4 | 64.5 | 66.2 | 69.50 |
| 7 | 75.1 | 74.3 | 73.9 | 73.8 | 70.9 | 70.6 | 70.5 | 70.6 | 70.8 | 70.6 | 70.3 | 71.5 | 70.45 |
| 8 | 73.5 | 72.5 | 71.5 | 71.6 | 71.1 | 71.1 | 71.5 | 71.5 | 71.2 | 71.1 | 70.9 | 70.6 | 70.63 |
| 9 | 76.1 | 76.1 | 75.0 | 73.6 | 72.3 | 71.1 | 71.6 | 70.9 | 71.0 | 71.5 | 70.7 | 70.9 | 71.66 |
| 10 | 76.1 | 76.9 | 75.0 | 72.3 | 70.6 | 70.1 | 70.5 | 70.9 | 70.9 | 71.0 | 71.1 | 70.8 | 70.98 |
| 11 | 74.8 | 75.0 | 74.1 | 72.5 | 71.8 | 71.7 | 71.5 | 71.2 | 71.1 | 70.9 | 70.8 | 70.6 | 71.00 |
| 12 | 76.6 | 76.7 | 75.7 | 74.2 | 72.0 | 71.8 | 71.1 | 71.0 | 70.5 | 70.1 | 68.0 | 70.0 | 71.36 |
| 13 | 76.9 | 76.8 | 76.8 | 75.5 | 74.0 | 73.8 | 73.9 | 73.9 | 74.0 | 73.9 | 72.8 | 73.3 | 72.11 |
| 14 | 80.1 | 78.9 | 77.6 | 76.1 | 74.3 | 73.9 | 73.5 | 74.0 | 74.2 | 74.0 | 74.0 | 73.9 | 74.35 |
| 15 | 78.1 | 76.1 | 74.5 | 72.6 | 70.9 | 70.2 | 70.5 | 71.1 | 71.3 | 71.2 | 71.4 | 71.1 | 72.88 |
| 16 | 76.8 | 77.0 | ... | 74.0 | 72.0 | 70.1 | 70.2 | 71.1 | 70.6 | 71.5 | 71.1 | 71.1 | 70.84 |
| 17 | 75.9 | 75.9 | 75.0 | 73.0 | 71.8 | 71.1 | 70.7 | 71.1 | 71.3 | 71.1 | 71.1 | 70.7 | 71.30 |
| 18 | 74.1 | 74.2 | 73.2 | 72.2 | 70.9 | 71.0 | 71.0 | 71.2 | 71.0 | 70.6 | 71.2 | 70.3 | 70.42 |
| 19 | 75.8 | 75.3 | 74.3 | 73.3 | 71.0 | 70.3 | 70.5 | 70.9 | 71.1 | 70.6 | 70.8 | 70.7 | 70.38 |
| 20 | 74.8 | 74.9 | 73.6 | 72.7 | 71.1 | 71.0 | 71.5 | 71.2 | 71.5 | 71.2 | 70.0 | 70.9 | 70.89 |
| 21 | 75.0 | 74.1 | 73.2 | 72.8 | 70.6 | 70.1 | 70.2 | 70.6 | 70.6 | 68.3 | 70.5 | 65.0 | 70.46 |
| 22 | 76.5 | 76.2 | 76.3 | 74.5 | 72.4 | 72.3 | ... | 71.0 | 71.2 | 70.9 | 70.6 | 71.0 | 71.71 |
| 23 | 76.2 | 74.9 | 73.3 | 71.2 | 69.6 | 69.5 | 70.0 | 69.4 | 71.2 | 71.5 | 71.0 | 71.5 | 70.76 |
| 24 | 75.1 | 74.6 | 73.2 | 72.0 | 70.6 | 70.5 | 71.0 | 71.0 | 71.5 | 71.5 | 71.4 | 71.0 | 70.34 |
| 25 | 74.5 | 73.1 | 72.6 | 71.2 | 70.1 | 70.5 | 71.1 | 71.6 | 71.6 | 71.5 | 71.2 | 70.6 | 70.98 |
| 26 | ... | 76.6 | 75.1 | 77.0 | 71.5 | 71.5 | 70.9 | 71.2 | 67.9 | 68.2 | 69.9 | 70.4 | 69.84 |
| 27 | 74.8 | 75.4 | 75.3 | 74.3 | 73.1 | 72.3 | 71.9 | 71.7 | 71.0 | 70.6 | 71.1 | 71.0 | 71.66 |
| 28 | 77.4 | 76.5 | 75.1 | 74.1 | 72.2 | 71.2 | 71.0 | 71.0 | 71.2 | 71.5 | 71.5 | 71.0 | 71.81 |
| 29 | 76.4 | 76.3 | 76.5 | 75.0 | 73.3 | 72.2 | 71.6 | 71.1 | 70.5 | 71.0 | 70.1 | 70.8 | 71.44 |
| 30 | 75.3 | 75.5 | 75.1 | 74.6 | 74.3 | 75.5 | 74.0 | 72.6 | 71.0 | 61.0 | 68.6 | 71.8 | 71.07 |
| 31 | 75.1 | 75.0 | 75.0 | 73.0 | 71.4 | 70.9 | 71.0 | 71.6 | 71.3 | 71.0 | 71.0 | 71.0 | 70.74 |

TABLE V—Continued—*Dec*

Ordinates, expressed in minutes of arc, taken from the daily declination traces. The ordinate

Base-line value

| Day. | 1. | 2. | 3. | 4. | 5. | 6. | 7. | 8. | 9. | 10. | 11. | 12. |
|---|---|---|---|---|---|---|---|---|---|---|---|---|
| 1 | 70.4 | 71.6 | 72.0 | 73.9 | 69.4 | 66.9 | 64.9 | 64.6 | 67.0 | 69.8 | 72.9 | ... |
| 2 | 71.1 | 70.6 | 70.1 | 70.5 | 68.5 | 67.5 | 65.2 | 66.9 | 67.0 | 68.9 | 72.0 | 74.8 |
| 3 | 71.5 | 70.5 | 70.3 | 69.5 | 68.5 | 67.9 | 67.4 | 67.1 | 67.9 | 71.9 | 73.5 | 74.0 |
| 4 | 70.5 | 70.5 | 70.4 | 69.9 | 69.2 | 68.1 | 66.9 | 66.2 | 67.2 | 69.5 | 73.0 | 75.6 |
| 5 | 71.0 | 70.5 | 70.2 | 69.9 | 69.2 | 67.6 | 67.1 | 67.4 | 68.0 | 69.9 | 73.6 | ... |
| 6 | 69.5 | 70.4 | 70.6 | 70.2 | 69.5 | 68.5 | 67.2 | 67.3 | 67.5 | 68.2 | 71.3 | 73.5 |
| 7 | 70.5 | 70.1 | 69.7 | 69.3 | 69.1 | 69.5 | 68.5 | 68.6 | 69.2 | 71.0 | 73.0 | 74.0 |
| 8 | 72.5 | 72.5 | 72.2 | 72.0 | 71.3 | 69.9 | 70.0 | 70.5 | 70.9 | ... | ... | ... |
| 9 | 67.9 | 71.8 | 71.5 | 76.1 | 68.4 | 69.5 | 66.9 | 67.6 | 70.0 | 73.0 | 73.0 | 74.0 |
| 10 | 71.1 | 70.8 | 70.1 | 70.1 | 69.2 | 68.4 | 67.4 | 67.0 | 67.3 | 68.9 | 71.3 | 72.7 |
| 11 | 69.6 | 73.0 | 70.2 | 69.5 | 68.5 | 67.5 | 67.1 | 67.1 | 67.6 | 68.5 | 69.9 | 72.0 |
| 12 | ... | ... | ... | ... | ... | ... | ... | ... | ... | ... | ... | ... |
| 13 | 70.0 | 70.0 | 69.9 | 69.2 | 68.2 | 67.2 | 67.9 | 65.5 | 66.2 | 69.0 | 73.1 | 75.6 |
| 14 | 68.8 | 70.5 | 72.2 | 70.3 | 78.6 | 72.4 | 73.9 | 78.1 | 78.0 | 82.4 | 77.1 | 82.5 |
| 15 | 68.4 | 71.3 | 71.6 | 71.3 | 69.3 | 67.0 | 65.9 | 66.5 | 67.9 | 70.4 | 73.0 | 75.3 |
| 16 | ... | ... | ... | ... | ... | ... | ... | ... | ... | ... | ... | ... |
| 17 | 70.0 | 69.9 | 70.0 | 69.1 | 68.9 | 67.1 | 65.6 | 65.5 | 77.4 | 69.0 | 71.3 | 74.1 |
| 18 | 70.0 | 70.3 | 70.4 | 70.1 | 68.3 | 66.8 | 66.0 | 66.1 | 67.7 | 70.0 | 72.0 | 72.9 |
| 19 | 70.3 | 69.9 | 70.5 | 70.3 | 69.0 | 68.2 | 67.0 | 67.3 | 70.1 | 72.5 | 74.0 | 74.6 |
| 20 | 70.5 | 70.4 | 70.0 | 69.5 | 68.7 | 66.6 | 64.9 | 64.5 | 67.0 | 69.5 | 71.3 | 73.2 |
| 21 | 67.5 | 69.0 | 70.6 | 66.1 | 66.1 | 65.3 | 65.6 | 65.2 | 67.0 | 71.6 | 72.0 | 73.8 |
| 22 | 72.8 | 69.0 | 69.7 | 69.0 | 68.4 | 69.0 | 66.5 | 66.5 | 68.6 | 70.2 | 70.0 | 75.3 |
| 23 | 70.5 | 70.1 | 71.9 | 71.5 | 70.6 | 66.8 | 66.5 | 67.5 | 68.1 | 70.5 | 72.5 | 74.1 |
| 24 | 70.8 | 70.3 | 70.1 | 70.0 | 69.4 | 67.5 | 65.9 | 65.9 | 67.2 | 71.0 | 73.9 | 74.9 |
| 25 | 71.2 | 70.1 | 69.9 | 70.1 | 69.1 | 67.5 | 66.9 | 66.2 | 66.9 | 70.0 | 73.0 | 75.2 |
| 26 | 70.9 | 71.0 | 70.5 | 69.5 | 68.5 | 66.9 | 65.2 | 65.6 | 67.5 | 71.5 | 74.5 | 76.1 |
| 27 | 71.0 | 70.6 | 70.1 | 69.9 | 69.0 | 66.5 | 65.1 | 64.9 | 66.6 | 71.0 | 74.7 | 76.2 |
| 28 | 69.7 | 69.5 | 69.0 | 68.5 | 66.9 | 65.0 | 62.4 | 62.0 | 62.1 | 66.0 | ... | ... |
| 29 | 69.6 | 71.0 | 70.4 | 69.6 | 68.1 | 66.3 | 64.5 | 63.6 | 63.2 | 67.0 | 71.0 | 74.9 |
| 30 | 70.5 | 69.9 | 70.9 | 69.2 | 68.8 | 67.2 | 65.4 | 64.9 | 66.3 | 70.0 | 72.5 | 74.5 |

*lination, June,* 1889.

for any hour added to the base-line value gives the absolute westerly declination at that hour.

−2° 49' 47''

| Day. | P. M. | | | | | | | | | | | | Mean. |
|---|---|---|---|---|---|---|---|---|---|---|---|---|---|
| | 1. | 2. | 3. | 4. | 5. | 6. | 7. | 8. | 9. | 10. | 11. | 12. | |
| 1 | 76.1 | 76.4 | 75.2 | 73.6 | 72.2 | 70.9 | 70.6 | 70.6 | 69.5 | 65.9 | 69.0 | 70.2 | 70.59 |
| 2 | 75.5 | 75.7 | 75.5 | 74.3 | 72.5 | 70.9 | 70.6 | 70.9 | 70.2 | 70.0 | 69.6 | 71.0 | 70.83 |
| 3 | 75.1 | 76.1 | 75.1 | 73.5 | 72.9 | 72.0 | 71.5 | 71.4 | 71.2 | 69.6 | 68.9 | 69.6 | 71.12 |
| 4 | 78.3 | 78.0 | 76.6 | 74.4 | 72.1 | 70.6 | 70.5 | 70.9 | 71.0 | 71.1 | 70.5 | 70.9 | 71.33 |
| 5 | 77.5 | 76.6 | 74.5 | 73.2 | 72.2 | 71.8 | 71.6 | 71.5 | 71.1 | 71.4 | 70.7 | 70.1 | 71.16 |
| 6 | 73.6 | 73.5 | 73.0 | 73.1 | 73.0 | 72.3 | 70.6 | 70.9 | 70.9 | 70.9 | 70.0 | 70.8 | 70.68 |
| 7 | 75.1 | 74.7 | 73.7 | 73.2 | 73.0 | 73.2 | 73.5 | 73.2 | 71.6 | 72.2 | 73.0 | 72.7 | 71.73 |
| 8 | 77.0 | 76.7 | 76.2 | 75.1 | 75.6 | 74.7 | 74.4 | 74.1 | 73.9 | 73.6 | 73.2 | 72.0 | 73.25 |
| 9 | 73.2 | 73.9 | 73.5 | 72.5 | 71.9 | 71.0 | 70.5 | 70.5 | 69.8 | 71.3 | 71.2 | 71.1 | 71.25 |
| 10 | 73.1 | 73.8 | 73.8 | 72.9 | 71.8 | 71.6 | 71.5 | 71.0 | 70.5 | 69.2 | 69.5 | 70.6 | 70.15 |
| 11 | 73.5 | 73.9 | 73.8 | 72.6 | 72.0 | 71.1 | 70.6 | ... | ... | ... | ... | ... | 70.42 |
| 12 | ... | ... | ... | ... | 72.1 | 70.5 | 71.0 | 70.4 | 70.2 | 70.6 | 71.0 | 70.6 | 70.80 |
| 13 | 82.6 | 80.9 | 80.0 | 78.9 | 77.7 | 77.6 | 77.0 | 76.6 | 76.3 | 76.0 | 73.5 | 72.9 | 73.41 |
| 14 | 75.3 | 75.2 | 74.5 | 74.3 | 72.7 | 71.5 | 73.0 | 71.9 | 70.9 | 71.2 | 72.4 | 70.6 | 74.10 |
| 15 | ... | ... | ... | ... | ... | ... | ... | ... | ... | ... | ... | ... | 69.82 |
| 16 | 71.6 | 72.2 | 75.4 | 75.5 | 73.9 | 72.1 | 71.2 | 71.0 | 67.3 | 69.5 | 70.5 | 70.6 | 71.73 |
| 17 | ... | ... | ... | ... | ... | 70.3 | 70.0 | 69.5 | 70.2 | 70.0 | 70.0 | 70.0 | 69.36 |
| 18 | 75.0 | 74.2 | 73.9 | 73.0 | 72.5 | 71.8 | 71.1 | 70.9 | 70.7 | 70.1 | 70.0 | 70.2 | 70.58 |
| 19 | 74.9 | 74.0 | 73.4 | 73.1 | 72.8 | 71.6 | 70.4 | 70.5 | 70.6 | 69.6 | 70.1 | 70.5 | 71.47 |
| 20 | 74.0 | 74.0 | 74.5 | 73.4 | 72.9 | 72.0 | 71.3 | 71.1 | 71.1 | 68.1 | 68.5 | 67.1 | 70.02 |
| 21 | ... | 73.0 | 71.5 | 71.4 | 72.0 | 71.1 | 71.1 | 69.1 | 67.7 | 69.2 | 69.5 | 70.5 | 69.39 |
| 22 | 74.4 | 73.4 | 73.5 | 72.5 | 72.5 | 72.0 | 71.5 | 71.4 | 71.5 | 71.1 | 71.4 | 71.0 | 70.88 |
| 23 | 73.9 | 73.6 | 73.4 | 72.6 | 71.7 | 71.0 | 71.0 | 71.0 | '69.0 | 70.4 | 70.6 | 70.6 | 70.81 |
| 24 | 74.5 | 74.0 | 74.4 | 72.6 | 70.9 | 70.6 | 71.1 | 71.2 | 71.1 | 70.2 | 70.0 | 70.6 | 70.75 |
| 25 | 76.7 | 76.1 | 74.9 | 73.5 | 72.1 | 71.1 | 71.5 | 71.2 | 69.1 | 70.3 | 70.6 | 71.0 | 71.01 |
| 26 | 76.5 | 76.3 | 75.4 | 73.3 | 71.6 | 70.6 | 70.5 | 70.5 | 70.6 | 70.6 | 70.6 | 70.7 | 71.04 |
| 27 | 76.4 | 74.7 | 74.0 | 72.5 | 71.3 | 70.8 | 71.1 | 70.7 | 70.9 | 70.9 | 70.7 | 69.6 | 70.80 |
| 28 | 77.9 | 78.4 | 75.5 | 76.4 | 74.3 | 71.8 | 68.6 | 68.0 | 69.4 | 69.6 | 69.4 | 69.9 | 69.56 |
| 29 | 76.8 | 76.2 | 74.9 | 72.9 | 71.9 | 71.4 | 70.4 | 70.6 | 70.6 | 70.1 | 70.5 | 70.5 | 70.25 |
| 30 | 75.2 | 74.8 | 73.0 | 71.9 | 70.7 | 70.0 | 70.5 | 70.6 | 70.8 | 68.8 | 70.6 | 70.9 | 70.75 |

MAGNETIC OBSERVATIONS AT THE U. S. NAVAL OBSERVATORY.

TABLE V—Continued.—*Dec*

Ordinates, expressed in minutes of arc, taken from the daily declination traces. The ordinate

Base-line value

| Day. | A. M. | | | | | | | | | | | |
|------|------|------|------|------|------|------|------|------|------|------|------|------|
| | 1. | 2. | 3. | 4. | 5. | 6. | 7. | 8. | 9. | 10. | 11. | 12. |
| 1 | 70. 1 | 68. 5 | 71. 2 | 66. 6 | 66. 1 | 64. 2 | 63. 0 | 63. 5 | 65. 2 | 68. 0 | 72. 8 | 74. 9 |
| 2 | 71. 5 | 70. 5 | 70. 4 | 69. 5 | 70. 4 | 69. 5 | 65. 6 | 65. 1 | 65. 5 | 67. 2 | 70. 6 | 72. 1 |
| 3 | 68. 2 | 70. 1 | 70. 6 | 70. 0 | 72. 1 | 68. 0 | 66. 4 | 67. 3 | 67. 2 | 68. 5 | 72. 0 | 75. 0 |
| 4 | 70. 5 | 70 5 | 70. 5 | 70. 5 | 69. 9 | 67. 4 | 66. 6 | 67. 2 | 67. 3 | 69. 4 | 71. 6 | 72. 0 |
| 5 | 71. 5 | 68. 3 | 68. 6 | 69. 4 | 69. 0 | 67. 3 | 66. 7 | 67. 0 | 68. 7 | 70. 1 | 72. 1 | 74. 0 |
| 6 | 68. 7 | 70. 6 | 70. 1 | 69. 3 | 68. 9 | 70. 3 | 66. 7 | 66. 7 | 66. 7 | 70. 7 | 74. 0 | 78. 1 |
| 7 | 69. 7 | 70. 0 | 72. 0 | 71. 5 | 70. 5 | 70. 1 | 70. 0 | 68 6 | 68. 8 | 71. 4 | 72. 6 | . . . |
| 8 | . . . | . . . | . . . | . . . | . . . | . . . | . . . | . . . | . . . | . . . | 70. 0 | 72. 0 |
| 9 | 71. 0 | 70. 1 | 71. 0 | 70. 5 | 69. 4 | 69. 4 | 67. 1 | 66. 6 | 68. 0 | 69. 8 | 71. 4 | 72. 5 |
| 10 | 71. 2 | 71. 2 | 70. 9 | 70. 9 | 70. 4 | 68. 9 | 67. 4 | 66. 6 | 67. 9 | 70. 9 | 73. 9 | 74. 9 |
| 11 | 71. 2 | 71. 0 | 70. 2 | 69. 8 | 68. 3 | 66. 5 | 64. 5 | 64. 1 | 64. 6 | 66. 9 | 69. 9 | 73. 5 |
| 12 | 71. 1 | 71. 1 | 70. 8 | 70. 2 | 68. 9 | 68. 0 | 66. 4 | 66. 4 | 67. 6 | 71. 1 | 72. 4 | 75. 2 |
| 13 | 71. 4 | 70. 8 | 70. 4 | 69. 9 | 69. 1 | 67. 6 | 66. 1 | 65. 0 | 65. 5 | 69. 9 | 72. 6 | 77. 4 |
| 14 | 71. 1 | 71. 1 | 70. 8 | 70. 1 | 69. 4 | 67. 4 | 66. 1 | 65. 6 | 66. 9 | 69. 1 | 72. 1 | 74. 0 |
| 15 | 69. 7 | 70. 2 | 70. 9 | 70. 0 | 69. 3 | 67. 9 | 67. 0· | 65. 7 | 65. 5 | 67. 0 | 69. 6 | 72. 1 |
| 16 | 70. 5 | 71. 0 | 71. 1 | 70. 7 | 70. 5 | 69. 9 | 68. 5 | 68. 8 | 69. 7 | 71. 1 | 73. 5 | 75. 1 |
| 17 | 69. 1 | 64. 2 | 50. 0 | 64. 0 | 65. 9 | 65. 5 | 68. 1 | 69. 9 | 68. 0 | 71. 9 | 71. 1 | 74. 0 |
| 18 | 70. 0 | 73. 4 | 70. 4 | 72. 6 | 69. 5. | 69. 4 | 68. 1 | 71. 2 | 70. 7 | 70. 0 | 71. 6 | 71. 6 |
| 19 | 72. 3 | 70. 8 | 68. 6 | 73. 0 | 70. 2 | 67. 7 | 67. 2 | 68. 9 | 67. 1 | 68. 7 | 70. 1 | 72. 0 |
| 20 | 74. 2 | 70. 8 | 71. 4 | 70. 9 | 71. 0 | 69. 4 | 68. 4 | 69. 2 | 70. 5 | 72. 6 | 74. 0 | 74. 4 |
| 21 | 71. 7 | 72. 2 | 72. 8 | 72. 3 | 71. 3 | 70. 0 | 68. 8 | 68. 6 | 69. 8 | 72. 7 | 74. 3 | 76. 1 |
| 22 | 73. 0 | 72. 6 | 72. 4 | 73. 0 | 72. 0 | 70. 9 | 69. 8 | 69. 8 | 70. 1 | 72. 2 | 74. 5 | 76. 5 |
| 23 | 72. 5 | 71. 6 | 71. 9 | 70. 8 | 70. 4 | 69. 1 | 68. 5 | 68. 5 | 68. 5 | 70. 1 | 72. 1 | 73. 5 |
| 24 | 71. 5 | 71. 6 | 71. 4 | 70. 9 | 69. 7 | 68. 3 | 67. 1 | 65. 4 | 67. 6 | 71. 0 | 75. 0 | 77. 0 |
| 25 | 72. 4 | 71. 6 | 71. 0 | 70. 6 | 69. 7 | 68. 0 | 66. 0 | 66. 7 | 66. 7 | 68. 8 | 72. 8 | 77. 6 |
| 26 | 71. 1 | 72. 0 | 71. 5 | 71. 2 | 72. 1 | 68. 9 | 66. 9 | 66. 6 | 67. 0 | 70. 1 | 73. 6 | 76. 0 |
| 27 | 72. 1 | 71. 8 | 71. 5 | 71. 5 | 70. 4 | 69. 0 | 68. 0 | 68. 2 | 68. 4 | 70. 8 | 73. 1 | 75. 5 |
| 28 | 71. 7 | 71. 4 | 71. 2 | 71. 1 | 71. 0 | 69. 8 | 69. 4 | 69. 4 | 70. 1 | 70. 7 | 72. 1 | 73. 5 |
| 29 | 70. 5 | 70. 7 | 69. 5 | 70. 1 | 71. 6 | 69. 6 | 67. 1 | 66. 8 | 67. 3 | 70. 0 | 70. 6 | 73. 5 |
| 30 | 71. 2 | 70. 9 | 76. 9 | 74. 5 | 70. 6 | 68. 4 | 67. 4 | 67. 3 | 67. 2 | 68. 6 | 71. 6 | 74. 0 |
| 31 | 64. 0 | 66. 0 | 71. 0 | 72. 0 | 71. 3 | 68. 0 | 66. 0 | 66. 6 | 68. 2 | 70. 0 | 72. 5 | 75. 5 |

*liuation, July,* 1889.

for any hour added to the base-line value gives the absolute westerly declination at that hour.

$= 2°\ 49'\ 15''$

| Day. | | | | | | P. M. | | | | | | | Mean. |
|---|---|---|---|---|---|---|---|---|---|---|---|---|---|
| | 1. | 2. | 3. | 4. | 5. | 6. | 7. | 8. | 9. | 10. | 11. | 12. | |
| 1 | 76.9 | 78.9 | 77.3 | 74.0 | 72.9 | 71.5 | 69.9 | 71.0 | 71.0 | 69.5 | 67.6 | 67.6 | 70.09 |
| 2 | 75.0 | 75.2 | 74.5 | 73.5 | 72.2 | 71.3 | 71.1 | 71.5 | 71.4 | 70.3 | 68.0 | 69.0 | 70.45 |
| 3 | 76.3 | 75.6 | 75.3 | 74.0 | 72.2 | 71.5 | 71.1 | 71.0 | 71.0 | 70.8 | 70.7 | 70.6 | 71.06 |
| 4 | 73.4 | 73.7 | 72.9 | 71.6 | 70.7 | 70.6 | 71.3 | 71.5 | 71.4 | 71.5 | 71.4 | 71.0 | 70.60 |
| 5 | . . . | 74.8 | 74.0 | 72.9 | 71.1 | 70.9 | 71.8 | 71.2 | 71.9 | 71.7 | 70.2 | 67.3 | 70.46 |
| 6 | 77.4 | 77.7 | 75.2 | 73.2 | 71.8 | 70.8 | 71.3 | 70.7 | 62.0 | 65.9 | 69.0 | 70.5 | 70.68 |
| 7 | 76.7 | 76.4 | . . . | . . . | . . . | . . . | | | | | | | 71.41 |
| 8 | 73.6 | 74.1 | 73.5 | 72.6 | 71.9 | 72.1 | 71.6 | 71.3 | 71.0 | 70.9 | 71.1 | 71.2 | 71.92 |
| 9 | 74.2 | 74.6 | 73.2 | 71.7 | 71.1 | 71.1 | 71.2 | 71.4 | 72.0 | 71.2 | 70.8 | 71.2 | 70.85 |
| 10 | 74.7 | 73.4 | 72.2 | 71.5 | 70.5 | 70.2 | 70.5 | 70.9 | 71.5 | 71.5 | 71.5 | 71.5 | 71.04 |
| 11 | 78.3 | 76.6 | 76.0 | 72.3 | 71.7 | 71.1 | 71.6 | 72.1 | 72.1 | 71.5 | 71.2 | 70.5 | 70.65 |
| 12 | 75.4 | 75.9 | 75.5 | 73.6 | 71.6 | 70.8 | 71.0 | 71.8 | 71.8 | 71.1 | 71.2 | 71.4 | 71.26 |
| 13 | 79.4 | 78.1 | 75.6 | 73.8 | 72.6 | 71.6 | 71.0 | 71.8 | 72.1 | 71.9 | 71.4 | 71.5 | 71.52 |
| 14 | 76.2 | 76.6 | 76.7 | 73.4 | 71.5 | 70.7 | 71.6 | 72.6 | 72.2 | 70.4 | 70.5 | 70.2 | 71.10 |
| 15 | 73.5 | 74.1 | 74.6 | 74.0 | 73.0 | 72.5 | 72.0 | 72.2 | 72.2 | 72.0 | 70.2 | 69.9 | 71.05 |
| 16 | 77.2 | 77.2 | 75.9 | 74.1 | 73.0 | 72.5 | 72.9 | 72.5 | 72.3 | 72.1 | 71.9 | 71.0 | 72.21 |
| 17 | 74.2 | 75.8 | 74.5 | 72.6 | 72.7 | 72.3 | 72.4 | 69.5 | 72.0 | 71.0 | 67.9 | 69.2 | 70.25 |
| 18 | 73.2 | 73.9 | 74.0 | 74.0 | 74.2 | 73.3 | 73.1 | 72.0 | 66.3 | 71.2 | 67.4 | 69.8 | 71.29 |
| 19 | 74.5 | 75.7 | 75.6 | 75.1 | 73.9 | 73.4 | 71.9 | 71.8 | 71.3 | 70.6 | 70.3 | 70.7 | 71.31 |
| 20 | 74.5 | 76.5 | 76.8 | 75.0 | 73.4 | 72.0 | 70.5 | 70.0 | 68.7 | 67.5 | 74.0 | 68.2 | 71.83 |
| 21 | 77.2 | 76.9 | 75.6 | 74.7 | 74.0 | 72.7 | 72.2 | 72.8 | 72.4 | 72.2 | 72.6 | 72.4 | 72.76 |
| 22 | 76.7 | 76.8 | 76.1 | 75.0 | 73.7 | 72.5 | 72.1 | 72.6 | 72.3 | 72.6 | 72.3 | 72.5 | 73.00 |
| 23 | 73.2 | 73.3 | 73.6 | 73.0 | 72.9 | 71.6 | 72.5 | 73.1 | 72.7 | 72.3 | 70.5 | 70.9 | 71.55 |
| 24 | 77.4 | 76.0 | 75.1 | 75.2 | 73.8 | 72.7 | 72.1 | 72.5 | 71.1 | 70.8 | 72.5 | 72.3 | 72.00 |
| 25 | 77.4 | 78.1 | 77.5 | 76.2 | 75.4 | 74.5 | 71.8 | 72.1 | 72.1 | 71.7 | 70.5 | 70.7 | 72.08 |
| 26 | 76.9 | 77.5 | 76.5 | 74.2 | 73.1 | 72.9 | 72.9 | 73.7 | 72.5 | 69.5 | 71.4 | 70.8 | 72.04 |
| 27 | 76.7 | 76.6 | 75.2 | 74.2 | 73.4 | 72.7 | 72.6 | 72.4 | 72.6 | 71.5 | 72.0 | 71.7 | 72.16 |
| 28 | 75.7 | 76.2 | 76.0 | 75.7 | 74.2 | 73.0 | 71.0 | 72.2 | 71.3 | 69.5 | 68.5 | 71.0 | 71.90 |
| 29 | 76.3 | 77.7 | 77.7 | 76.7 | 74.5 | ·71.7 | 71.3 | 72.2 | 71.4 | 71.8 | 71.5 | 71.1 | 71.72 |
| 30 | 76.4 | 77.3 | 76.0 | 74.6 | 73.1 | 71.2 | 70.8 | 71.2 | 71.7 | 71.7 | 71.9 | 67.8 | 71.76 |
| 31 | 77.5 | 77.8 | 77.1 | 75.6 | 73.7 | 73.1 | 71.4 | 59.0 | 69.5 | 71.0 | 70.9 | 69.6 | 71.23 |

TABLE V—Continued—*Dec*

Ordinates, expressed in minutes of arc, taken from the daily declination traces. The ordinate

Base-line value

| Day. | | | | | | A. M. | | | | | | |
|---|---|---|---|---|---|---|---|---|---|---|---|---|
| | 1. | 2. | 3. | 4. | 5. | 6. | 7. | 8. | 9. | 10. | 11. | 12. |
| 1 | 72.2 | 72.3 | 70.8 | 72.8 | 70.6 | 68.5 | 67.0 | 66.0 | 67.1 | 69.9 | 72.0 | 74.1 |
| 2 | 71.8 | 69.6 | 72.2 | 73.9 | 73.1 | 69.1 | 68.1 | 67.0 | 67.0 | 69.4 | 73.5 | 78.4 |
| 3 | 72.6 | 72.2 | 72.0 | 72.6 | 72.3 | 70.8 | 69.4 | 67.5 | 68.8 | 71.8 | 74.2 | 75.2 |
| 4 | 72.0 | 72.1 | 71.8 | 71.2 | 71.4 | 69.3 | 68.2 | 67.6 | 69.9 | 73.0 | 75.6 | 76.4 |
| 5 | 71.7 | 71.6 | 71.6 | 72.1 | 71.4 | 69.9 | 67.8 | 67.1 | 67.8 | 69.9 | 72.1 | 74.0 |
| 6 | 71.4 | 72.0 | 72.1 | 71.7 | 71.6 | 70.0 | 69.0 | 68.6 | 70.0 | 73.5 | 76.1 | 77.1 |
| 7 | 72.1 | 72.0 | 71.6 | 71.6 | 71.0 | 69.9 | 68.5 | 68.2 | 69.5 | 73.2 | 76.1 | 77.1 |
| 8 | 76.2 | 70.4 | 71.0 | 71.1 | 70.2 | 70.0 | 66.8 | 67.0 | 68.4 | 71.5 | 74.3 | 77.5 |
| 9 | 72.4 | 72.0 | 71.7 | 71.2 | 70.8 | 68.8 | 66.7 | 65.6 | 66.2 | 71.1 | 74.7 | 77.0 |
| 10 | 72.1 | 72.0 | 71.4 | 70.8 | 69.5 | 67.3 | 66.0 | 65.7 | 68.0 | 72.8 | 76.0 | 77.7 |
| 11 | 71.7 | 69.7 | 68.7 | 68.3 | 68.1 | 64.3 | 62.0 | 61.5 | 64.9 | 68.9 | 73.0 | 76.3 |
| 12 | 70.3 | 71.9 | 71.2 | 70.0 | 69.1 | 66.2 | 65.2 | 64.6 | 65.8 | 69.4 | 73.0 | 75.4 |
| 13 | 66.5 | 69.5 | 72.6 | 71.3 | 70.1 | 68.0 | 70.1 | 67.9 | 69.6 | 73.7 | 75.6 | 76.1 |
| 14 | 71.9 | 73.9 | 72.5 | 71.2 | 72.0 | 70.6 | 67.2 | 66.6 | 67.8 | 70.5 | 73.2 | 75.3 |
| 15 | 71.7 | 72.0 | 70.7 | 70.9 | 72.2 | 68.0 | 71.4 | 69.1 | 68.8 | 71.7 | . . . | . . . |
| 16 | 71.5 | 73.9 | 75.8 | 72.8 | 71.1 | 68.9 | 68.3 | 67.4 | 68.4 | 71.2 | 73.3 | 74.4 |
| 17 | 72.6 | 72.4 | 72.8 | 71.6 | 71.9 | 69.7 | 68.1 | 66.6 | 68.0 | 71.6 | 74.5 | 76.6 |
| 18 | 72.7 | 72.2 | 74.0 | 71.5 | 69.9 | 69.3 | 67.7 | 67.4 | 69.4 | 73.1 | 76.6 | 78.0 |
| 19 | 72.5 | 73.3 | 74.0 | 71.8 | 70.0 | 68.5 | 69.1 | 71.7 | 73.7 | 75.1 | . . . | . . . |
| 20 | 70.7 | 69.1 | 70.2 | 68.9 | 70.4 | 68.9 | 73.8 | 70.0 | 72.9 | 73.2 | 75.5 | 78.7 |
| 21 | 70.4 | 71.7 | 71.7 | 71.2 | 70.2 | 68.9 | 66.7 | 66.5 | 68.5 | 72.2 | 75.1 | 76.1 |
| 22 | 72.6 | 72.2 | 71.7 | 71.2 | 70.4 | 69.7 | 68.1 | 68.2 | 70.6 | 73.0 | 75.9 | 77.3 |
| 23 | 72.7 | 72.2 | 72.1 | 71.5 | 71.0 | 68.4 | 66.0 | 65.5 | 67.6 | 71.6 | 74.5 | 76.0 |
| 24 | 72.6 | 72.2 | 72.0 | 71.3 | 70.8 | 69.4 | 68.0 | 67.4 | 68.4 | 71.8 | 75.3 | 77.7 |
| 25 | 72.5 | 71.5 | 71.4 | 71.2 | 69.5 | 67.0 | 67.5 | 66.6 | 72.8 | 76.0 | 78.6 | 79.5 |
| 26 | 71.3 | 74.4 | 74.5 | 71.4 | 76.0 | 75.6 | 69.0 | 67.9 | 70.5 | 71.0 | 77.5 | 79.5 |
| 27 | 75.2 | 73.3 | 71.0 | 74.5 | 73.4 | 70.4 | 67.0 | 66.2 | 71.2 | 72.2 | 76.3 | 78.9 |
| 28 | 66.6 | 73.2 | 71.0 | 71.1 | 70.6 | 68.5 | 67.6 | 68.5 | 70.5 | 71.6 | 74.9 | 77.1 |
| 29 | 69.9 | 72.5 | 75.8 | 73.3 | 72.1 | 69.6 | 67.0 | 66.8 | 68.3 | 71.0 | 74.3 | 76.0 |
| 30 | 73.9 | 74.0 | 74.3 | 72.7 | 73.4 | 76.0 | 69.2 | 67.6 | 68.4 | 71.8 | 74.7 | 76.9 |
| 31 | 73.1 | 73.3 | 74.4 | 72.6 | 72.5 | 72.0 | 70.5 | 69.6 | 70.4 | 71.9 | 75.4 | 78.0 |

*lination, August, 1889.*

for any hour added to the base-line value gives the absolute westerly declination at that hour.

= 2° 48′ 55″

| Day. | P. M. | | | | | | | | | | | | Mean. |
|---|---|---|---|---|---|---|---|---|---|---|---|---|---|
| | 1. | 2. | 3. | 4. | 5. | 6. | 7. | 8. | 9. | 10. | 11. | 12. | |
| 1 | 73.5 | 77.4 | 76.3 | 77.0 | 73.9 | 69.5 | 72.2 | 72.1 | 67.8 | 69.5 | 68.9 | 69.7 | 71.30 |
| 2 | 80.0 | 81.3 | 80.4 | 77.8 | 73.6 | 72.7 | 71.8 | 71.9 | 72.2 | 72.1 | 72.1 | 73.4 | 73.02 |
| 3 | 75.2 | 75.0 | 74.3 | 73.0 | 71.7 | 71.1 | 71.6 | 71.8 | 71.7 | 71.5 | 71.5 | 71.0 | 72.03 |
| 4 | 76.5 | 76.5 | 74.9 | 73.2 | 72.4 | 72.0 | 72.1 | 72.4 | 72.4 | 71.7 | 71.7 | 71.5 | 72.33 |
| 5 | 74.9 | 75.3 | 74.4 | 72.7 | 72.1 | 72.3 | 72.6 | 72.7 | 72.5 | 72.3 | 71.3 | 72.0 | 71.75 |
| 6 | 76.9 | 76.4 | 75.4 | 74.3 | 73.2 | 72.5 | 72.5 | 72.6 | 72.6 | 72.8 | 72.5 | 72.0 | 72.78 |
| 7 | 78.0 | 77.2 | 76.0 | 74.5 | 73.4 | 73.2 | 73.3 | 73.0 | 73.1 | 72.6 | 72.6 | 72.6 | 72.93 |
| 8 | 79.2 | 77.8 | 76.1 | 74.3 | 73.2 | 71.9 | 72.0 | 72.3 | 72.1 | 71.4 | 72.4 | 72.2 | 72.47 |
| 9 | 78.3 | 78.3 | 77.0 | 74.7 | 73.2 | 72.0 | 71.6 | 71.6 | 70.8 | 71.8 | 72.2 | 70.7 | 72.10 |
| 10 | 78.5 | . . . | . . . | 74.0 | 72.7 | 72.0 | 72.0 | 72.2 | 73.0 | 73.8 | 73.2 | 73.1 | 71.99 |
| 11 | 78.2 | 78.2 | 76.9 | 74.8 | 73.0 | 71.8 | 71.5 | 72.0 | 71.5 | 71.7 | 70.0 | 71.5 | 70.77 |
| 12 | 77.6 | 77.7 | 76.4 | 74.4 | 72.9 | 72.4 | 73.0 | 73.5 | 72.2 | 69.0 | 71.8 | 73.1 | 71.50 |
| 13 | 79.3 | 78.3 | 78.1 | 75.0 | 73.8 | 66.9 | 72.8 | 73.8 | 73.4 | 72.8 | 71.6 | 71.3 | 72.42 |
| 14 | 76.2 | 76.2 | 75.4 | 74.3 | 73.2 | 72.3 | 72.3 | 72.6 | 72.0 | 72.2 | 72.6 | 71.7 | 72.24 |
| 15 | . . . | 76.0 | 75.8 | 73.7 | 73.1 | 73.2 | 72.0 | 72.7 | 70.6 | 69.4 | 68.1 | 69.6 | 71.46 |
| 16 | 75.4 | 75.1 | 74.4 | 73.5 | 73.6 | 73.5 | 73.1 | 73.0 | 72.9 | 72.6 | 72.7 | 72.7 | 72.48 |
| 17 | 77.5 | 76.9 | 75.4 | 73.3 | 72.0 | 71.7 | 71.7 | 72.0 | 72.5 | 72.4 | 72.4 | 72.5 | 72.36 |
| 18 | 77.5 | 74.9 | 73.4 | 72.5 | 73.0 | 72.9 | 73.0 | 72.8 | 72.7 | 72.5 | 72.3 | 72.6 | 72.58 |
| 19 | 76.9 | 76.4 | 74.7 | 73.3 | 72.5 | 72.0 | 72.1 | 72.1 | 72.4 | 72.4 | 72.1 | 72.1 | 72.67 |
| 20 | 78.2 | 77.0 | 77.6 | 78.0 | 76.1 | 75.0 | 73.3 | 72.5 | 69.6 | 71.7 | 72.9 | 70.6 | 73.12 |
| 21 | 78.1 | 77.8 | 75.4 | 74.4 | 73.5 | 73.1 | 73.0 | 71.8 | 73.3 | 73.2 | 73.1 | 72.9 | 72.45 |
| 22 | 77.8 | 77.4 | 75.4 | 74.0 | 72.8 | 72.6 | 72.5 | 73.5 | 73.1 | 72.7 | 71.9 | 72.1 | 72.78 |
| 23 | 77.6 | 77.5 | 76.1 | 73.9 | 72.2 | 72.3 | 73.2 | 73.3 | 71.4 | 70.6 | 72.4 | 72.0 | 72.15 |
| 24 | 78.4 | 78.0 | 76.0 | 74.2 | 73.3 | 72.0 | 72.6 | 72.5 | 72.8 | 72.9 | 72.8 | 70.3 | 72.58 |
| 25 | 80.0 | 79.5 | 76.1 | 73.9 | 72.4 | 72.3 | 72.8 | 72.5 | 72.8 | 71.8 | 73.0 | 71.4 | 73.02 |
| 26 | 80.5 | 78.6 | 76.6 | 74.5 | 72.0 | 71.7 | 71.4 | 71.9 | 69.0 | 70.5 | 68.5 | 70.1 | 73.08 |
| 27 | 80.0 | 79.2 | 76.4 | 74.5 | 70.0 | 72.5 | 72.3 | 71.7 | 70.5 | 71.1 | 71.5 | 66.6 | 72.75 |
| 28 | 78.8 | 77.6 | 76.5 | 74.8 | 73.5 | 72.1 | 71.0 | 69.7 | 70.1 | 69.0 | 72.5 | 70.0 | 71.95 |
| 29 | 77.5 | 77.6 | 76.9 | 75.0 | 73.6 | 73.3 | 72.9 | 70.6 | 71.3 | 70.6 | 72.9 | 73.4 | 72.59 |
| 30 | 78.2 | 78.2 | 77.0 | 75.4 | 74.4 | 73.7 | 73.1 | 73.1 | 73.1 | 73.0 | 73.0 | 72.5 | 73.65 |
| 31 | 78.8 | 78.7 | 77.7 | 76.0 | 74.2 | 73.4 | 73.6 | 73.7 | 73.6 | 73.5 | 73.2 | 73.5 | 73.90 |

TABLE V—Continued—*Dec*

Ordinates, expressed in minutes of arc, taken from the daily declination traces.  The ordinate

Base-line value

| Day | A. M. | | | | | | | | | | | |
|---|---|---|---|---|---|---|---|---|---|---|---|---|
| | 1. | 2. | 3. | 4. | 5. | 6. | 7. | 8. | 9. | 10. | 11. | 12. |
| 1 | 73.0 | 73.4 | 72.5 | 72.1 | 71.9 | 70.9 | 69.0 | 68.1 | 69.1 | 71.6 | 75.0 | 77.2 |
| 2 | 73.0 | 72.8 | 72.3 | 71.2 | 70.9 | 73.4 | 69.1 | 69.1 | 71.0 | 74.8 | 77.1 | 78.9 |
| 3 | 73.2 | 73.0 | 72.3 | 72.0 | 71.3 | 70.1 | 69.3 | 69.5 | 71.3 | 72.9 | 75.3 | 77.3 |
| 4 | 72.5 | 72.8 | 72.6 | 72.0 | 71.7 | 70.9 | 68.9 | 69.3 | 72.2 | 75.3 | 78.0 | 79.8 |
| 5 | 73.0 | 72.8 | 72.2 | 72.0 | 71.8 | 70.7 | 69.3 | 69.2* | 71.5 | 74.0 | 76.1 | 77.2 |
| 6 | 73.0 | 72.6 | 72.3 | 72.0 | 71.8 | 70.7 | 69.0 | 69.0 | 71.7 | 75.0 | 77.0 | 78.0 |
| 7 | 73.4 | 73.0 | 72.9 | 72.5 | 72.0 | 72.2 | 69.9 | 69.9 | 73.1 | 76.0 | 78.8 | 80.5 |
| 8 | 73.0 | 71.6 | 69.0 | 70.8 | 72.0 | 77.3 | 70.3 | 70.1 | 70.3 | 73.2 | 77.0 | 78.2 |
| 9 | 73.0 | 70.7 | 84.0 | 71.2 | 76.1 | 63.2 | 71.0 | 73.0 | 69.8 | 73.4 | 77.5 | 80.2 |
| 10 | 74.9 | 76.3 | 74.5 | 75.0 | 80.7 | 78.0 | 70.8 | 71.2 | 71.3 | 72.1 | 74.3 | 78.0 |
| 11 | 74.0 | 78.3 | 73.0 | 73.9 | 74.0 | 71.0 | 69.0 | 69.3 | 71.5 | 76.8 | 78.7 | 81.2 |
| 12 | 73.9 | 74.0 | 71.7 | 77.9 | 71.2 | 70.6 | 69.4 | 69.2 | 70.2 | 74.0 | 76.3 | 79.4 |
| 13 | 73.2 | 78.0 | 74.5 | 72.5 | 72.8 | 71.1 | 70.4 | 71.1 | 72.9 | 74.9 | 78.0 | 80.1 |
| 14 | 74.0 | 74.3 | 74.0 | 73.0 | 73.0 | 71.5 | 70.9 | 71.1 | 73.5 | 75.5 | 77.7 | 79.0 |
| 15 | 73.5 | 72.7 | 73.1 | 71.1 | 71.0 | 71.1 | 70.3 | 70.9 | 72.0 | 73.2 | 76.6 | 77.9 |
| 16 | 72.9 | 72.9 | 72.2 | 71.9 | 72.4 | 72.0 | 70.5 | 69.0 | 70.7 | 71.0 | 73.3 | 75.3 |
| 17 | 73.2 | 72.7 | 72.7 | 72.6 | 72.4 | 72.3 | 71.5 | 70.8 | 71.0 | 71.9 | 73.2 | 75.9 |
| 18 | 73.4 | 73.0 | 73.5 | 71.2 | 70.4 | 71.2 | 70.1 | 70.0 | 71.2 | 72.4 | 73.8 | 76.0 |
| 19 | 73.1 | 73.1 | 72.9 | 73.3 | 72.7 | 72.2 | 70.7 | 71.0 | 72.0 | 73.3 | 75.5 | 77.3 |
| 20 | 74.4 | 73.4 | 73.2 | 73.1 | 73.3 | 72.9 | 71.6 | 71.0 | 71.9 | 73.0 | 74.9 | 76.1 |
| 21 | 74.6 | 74.0 | 73.4 | 73.1 | 72.6 | 71.9 | 70.1 | 69.3 | 70.3 | 73.1 | 78.0 | 79.3 |
| 22 | 73.6 | 73.0 | 74.5 | 72.7 | 70.6 | 72.6 | 78.6 | 72.5 | 74.4 | 72.8 | 80.0 | 79.8 |
| 23 | 81.1 | 74.0 | 76.3 | 77.1 | 72.0 | 75.0 | 72.0 | 71.4 | 71.2 | 76.3 | 80.5 | 81.8 |
| 24 | 74.2 | 74.2 | 75.6 | 72.1 | 74.6 | 78.1 | 73.5 | 72.9 | 74.5 | 74.8 | 76.1 | 77.4 |
| 25 | 79.0 | 74.7 | 73.0 | 72.0 | 77.0 | 74.5 | 71.6 | 71.0 | 71.5 | 73.4 | 76.9 | 79.5 |
| 26 | 74.2 | 80.0 | 75.5 | 73.0 | 73.2 | 73.1 | 72.6 | 71.8 | 71.2 | 72.2 | 74.5 | 77.4 |
| 27 | 73.3 | 74.2 | 73.8 | 75.8 | 74.7 | 73.0 | 71.1 | 70.3 | 71.3 | 72.0 | 75.0 | 78.0 |
| 28 | 74.3 | 74.6 | 73.9 | 73.9 | 74.9 | 73.0 | 71.4 | 70.2 | 71.0 | 72.0 | 74.3 | 75.9 |
| 29 | 74.2 | 74.1 | 75.1 | 73.7 | 73.2 | 73.0 | 71.3 | 69.8 | 69.9 | 72.0 | 74.4 | 77.2 |
| 30 | 73.2 | 73.6 | 73.1 | 73.1 | 73.0 | 73.0 | 71.3 | 70.5 | 69.1 | 70.0 | 73.0 | 76.2 |

*lination, September,* 1889.

for any hour added to the base-line value gives the absolute westerly declination at that hour.

= 2° 47′ 44″

| Day. | P. M. | | | | | | | | | | | | Mean. |
|---|---|---|---|---|---|---|---|---|---|---|---|---|---|
| | 1. | 2. | 3. | 4. | 5. | 6. | 7. | 8. | 9. | 10. | 11. | 12. | |
| 1 | 78.8 | 78.5 | 77.1 | 75.1 | 73.8 | 73.2 | 73.8 | 73.9 | 73.8 | 73.5 | 733. | 73.1 | 73.40 |
| 2 | 79.2 | 78.3 | 76.7 | 75.0 | 74.1 | 73.8 | 72.6 | 74.0 | 74.4 | 74.0 | 74.3 | 73.6 | 73.90 |
| 3 | 77.7 | 77.3 | 76.5 | 75.0 | 73.5 | 73.7 | 74.1 | 74.2 | 74.0 | 73.8 | 73.6 | 73.2 | 73.50 |
| 4 | 79.5 | 78.2 | 76.2 | 74.5 | 74.0 | 74.0 | 74.0 | 74.0 | 74.0 | 73.1 | 73.5 | 73.2 | 73.92 |
| 5 | 77.1 | 76.2 | 74.2 | 73.0 | 73.0 | 73.5 | 74.0 | 73.9 | 73.2 | 74.0 | 73.7 | 73.1 | 73.28 |
| 6 | . . . | 77.4 | 75.6 | 74.2 | 73.0 | 74.5 | 74.0 | 74.2 | 74.1 | 74.0 | 73.9 | 73.5 | 73.50 |
| 7 | 80.5 | 79.0 | 76.2 | 73.8 | 74.0 | 75.7 | 76.0 | 74.5 | 74.0 | 74.0 | 73.4 | 73.4 | 74.53 |
| 8 | 79.1 | 77.0 | 75.7 | 74.9 | 75.0 | 72.0 | 72.0 | 72.0 | 71.0 | 71.8 | 67.3 | 76.8 | 73.23 |
| 9 | 80.8 | 78.1 | 77.4 | 72.3 | 74.2 | 73.8 | 59.0 | 68.0 | 70.6 | 71.1 | 73.1 | 79.5 | 73.38 |
| 10 | 80.0 | 80.1 | 78.8 | 76.5 | 74.1 | 72.9 | 59.0 | 67.4 | 72.2 | 70.1 | 74.0 | 73.0 | 73.97 |
| 11 | 81.2 | 80.1 | 78.2 | 75.7 | 74.1 | 72.6 | 69.4 | 71.1 | 73.7 | 71.5 | 71.0 | 73.0 | 74.26 |
| 12 | . . . | 79.9 | 77.4 | 75.3 | 73.6 | 73.1 | 73.7 | 72.6 | 73.8 | 74.1 | 73.1 | 74.2 | 73.85 |
| 13 | 80.1 | 78.5 | 77.0 | 75.3 | 74.1 | 73.6 | 73.5 | 73.7 | 73.9 | 74.0 | 74.0 | 74.4 | 74.65 |
| 14 | 78.7 | 78.1 | 76.8 | 75.0 | 73.0 | 73.2 | 74.0 | 73.8 | 73.9 | 73.9 | 73.4 | | 74.38 |
| 15 | 78.0 | 78.1 | 77.0 | 75.7 | 74.8 | 74.0 | 74.1 | 74.1 | 72.7 | 73.3 | 72.7 | 72.9 | 73.78 |
| 16 | 76.0 | 77.0 | 76.6 | 75.6 | 75.5 | 74.8 | 74.8 | 73.7 | 73.6 | 74.1 | 73.9 | 73.6 | 73.43 |
| 17 | 77.5 | 79.3 | 78.0 | 77.0 | 76.0 | 74.1 | 74.0 | 74.0 | 73.9 | 74.0 | 73.9 | 73.6 | 73.98 |
| 18 | 77.8 | 78.0 | 77.0 | 76.1 | 25.8 | 73.4 | 75.4 | 75.3 | 71.9 | 73.4 | 74.2 | 74.3 | 73.70 |
| 19 | 78.1 | 77.6 | 76.1 | 74.4 | 74.0 | 74.6 | 74.3 | 73.0 | 72.3 | 74.6 | 72.8 | 74.8 | 73.90 |
| 20 | . . . | . . . | . . . | . . . | . . . | | | | | | . . . | . . . | 73.23 |
| 21 | 79.6 | 78.2 | 76.5 | 74.6 | 73.8 | 74.5 | 74.0 | 69.7 | 73.4 | 74.1 | 74.6 | 74.0 | 74.03 |
| 22 | 80.9 | 79.6 | 78.1 | 77.7 | 75.6 | 70.6 | 71.1 | 73.0 | 71.0 | 71.6 | 75.4 | 74.8 | 74.32 |
| 23 | 78.8 | 78.0 | 76.0 | 73.8 | 71.1 | 74.2 | 74.2 | 73.0 | 72.0 | 73.4 | 73.0 | 73.5 | 74.99 |
| 24 | 77.6 | 77.1 | 76.9 | 75.8 | 72.4 | 74.5 | 74.0 | 68.0 | 72.6 | 73.9 | 73.6 | 74.2 | 74.53 |
| 25 | 79.0 | 78.8 | 77.0 | 75.0 | 73.5 | 73.3 | 73.9 | 74.2 | 72.0 | 73.1 | 72.5 | 73.5 | 74.58 |
| 26 | . . . | 77.1 | 75.6 | 74.1 | 73.9 | 74.0 | 73.9 | 74.0 | 74.0 | 71.8 | 72.9 | 73.9 | 74.08 |
| 27 | 78.0 | 78.0 | 77.0 | 75.5 | 74.3 | 73.9 | 73.8 | 73.7 | 73.9 | 72.8 | 73.7 | 75.0 | 74.25 |
| 28 | 76.1 | 76.5 | 76.1 | 75.2 | 75.1 | 74.1 | 74.0 | 74.0 | 74.0 | 74.3 | 74.3 | 74.1 | 74.05 |
| 29 | 77.6 | 77.7 | 76.7 | 75.3 | 75.1 | 75.0 | 74.1 | 73.9 | 74.1 | 74.1 | 74.2 | 73.3 | 74.13 |
| 30 | 77.5 | 77.4 | 86.7 | 75.9 | 75.6 | 75.1 | 74.5 | 74.4 | 74.3 | 74.2 | 73.3 | 73.6 | 73.82 |

TABLE V—Continued—*Dec*

Ordinates, expressed in minutes of arc, taken from the daily declination traces.   The ordinate

Base-line value

| Day. | -A.M. | | | | | | | | | | | |
|---|---|---|---|---|---|---|---|---|---|---|---|---|
| | 1. | 2. | 3. | 4. | 5. | 6. | 7. | 8. | 9. | 10. | 11. | 12. |
| 1 | 74.0 | 73.8 | 73.7 | 73.3 | 73.2 | 73.3 | 71.7 | 70.0 | 69.4 | 70.9 | 73.1 | 76.1 |
| 2 | 74.0 | 73.6 | 73.6 | 73.0 | 73.0 | 72.3 | 71.3 | 70.7 | 70.3 | 71.5 | 74.0 | 77.0 |
| 3 | 73.5 | 69.5 | 68.8 | 70.0 | 72.5 | 72.3 | 70.2 | 70.1 | 71.1 | 72.3 | 73.8 | 75.4 |
| 4 | 73.9 | 73.1 | 73.1 | 73.2 | 73.8 | 73.3 | 73.0 | 72.0 | 71.4 | 71.6 | 73.0 | 75.4 |
| 5 | 74.1 | 74.0 | 74.0 | 73.3 | 73.1 | 73.5 | 74.2 | 74.6 | 73.8 | 74.5 | 76.0 | 79.9 |
| 6 | 73.9 | 72.1 | 77.0 | 77.1 | 74.2 | 74.9 | 72.3 | 70.9 | 72.0 | 74.0 | 75.7 | 78.5 |
| 7 | 75.3 | 78.1 | 74.9 | 72.7 | 74.2 | 72.8 | 71.5 | 74.5 | 75.0 | 75.0 | 76.8 | 77.5 |
| 8 | 75.0 | 73.7 | 79.7 | 74.5 | 72.9 | 77.3 | 75.1 | 71.6 | 72.9 | 75.4 | 77.5 | 79.0 |
| 9 | 77.2 | 76.0 | 74.0 | 77.3 | 75.3 | 72.9 | 73.6 | 73.2 | 73.1 | 75.1 | 78.4 | 79.3 |
| 10 | 74.1 | 76.2 | 74.8 | 73.4 | 73.2 | 73.0 | 73.0 | 72.6 | 73.9 | 76.2 | 79.3 | 80.1 |
| 11 | 74.5 | 74.3 | 74.2 | 74.0 | 73.7 | 73.1 | 71.9 | 71.3 | 71.3 | 73.1 | 76.6 | 79.0 |
| 12 | 74.7 | 74.2 | 74.1 | 73.4 | 73.6 | 73.3 | 72.0 | 70.6 | 70.9 | 72.0 | 76.2 | 77.5 |
| 13 | 74.5 | 74.0 | 74.0 | 73.8 | 73.0 | 74.2 | 73.0 | 71.0 | 71.2 | 73.2 | 76.3 | 78.0 |
| 14 | 74.0 | 71.9 | 74.1 | 72.0 | 73.0 | 73.0 | 71.7 | 71.2 | 70.3 | 71.5 | 73.4 | 75.9 |
| 15 | 74.7 | 75.1 | 74.9 | 74.7 | 74.0 | 73.9 | 72.8 | 71.0 | 70.0 | 70.8 | 73.0 | 76.9 |
| 16 | 74.0 | 77.2 | 72.3 | 73.5 | 74.0 | 73.6 | 72.5 | 71.5 | 71.1 | 71.8 | 73.3 | 75.3 |
| 17 | 74.8 | 74.3 | 74.6 | 74.3 | 74.2 | 73.6 | 72.2 | 72.8 | 71.3 | 72.4 | 75.4 | 78.4 |
| 18 | 74.4 | 75.2 | 74.6 | 74.2 | 74.0 | 74.2 | 72.7 | 71.0 | 70.0 | 71.8 | 74.7 | 78.4 |
| 19 | 75.3 | 75.9 | 75.1 | 74.6 | 74.3 | 76.4 | 74.7 | 71.0 | 70.7 | 75.0 | 77.9 | 79.2 |
| 20 | 73.5 | 74.2 | 76.0 | 80.0 | 74.7 | 79.5 | 77.6 | 75.1 | 71.9 | 73.7 | 78.6 | 80.0 |
| 21 | 70.7 | 76.1 | 74.3 | 74.2 | 75.7 | 74.4 | 74.3 | 72.3 | 72.5 | 73.0 | 76.1 | 77.0 |
| 22 | 74.8 | 76.9 | 76.0 | 75.7 | 76.9 | 74.3 | 74.7 | 73.0 | 74.2 | 74.4 | 75.7 | 76.3 |
| 23 | 75.0 | 75.9 | 75.4 | 74.1 | 74.4 | 76.2 | 73.3 | 74.4 | 75.4 | 74.2 | 75.3 | 76.2 |
| 24 | 75.2 | 74.9 | 74.6 | 74.3 | 74.3 | 74.1 | 73.1 | 71.8 | 72.6 | 74.0 | 76.4 | 77.4 |
| 25 | 75.1 | 75.0 | 74.8 | 74.2 | 75.2 | 74.4 | 73.1 | 72.6 | 73.1 | 74.1 | 75.6 | 77.5 |
| 26 | 74.9 | 74.7 | 74.4 | 74.8 | 73.2 | 73.4 | 71.6 | 71.3 | 72.0 | 74.1 | 76.2 | 78.0 |
| 27 | 74.2 | 73.8 | 74.3 | 73.9 | 73.9 | 73.5 | 72.3 | 70.9 | 70.7 | 72.0 | 74.7 | 77.0 |
| 28 | 74.5 | 73.7 | 74.7 | 74.5 | 73.7 | 73.4 | 72.5 | 70.8 | 71.4 | 74.0 | 77.5 | 79.3 |
| 29 | 74.8 | 74.5 | 74.3 | 74.1 | 74.3 | 73.9 | 73.0 | 72.0 | 71.4 | 71.9 | 73.1 | 74.9 |
| 30 | 74.7 | 74.8 | 75.7 | 73.9 | 73.5 | 73.8 | 72.9 | 72.1 | 72.0 | 72.3 | 72.7 | 75.5 |
| 31 | 75.0 | 74.3 | 74.4 | 74.5 | 74.1 | 73.6 | 73.4 | 72.5 | 72.1 | . . . | 74.7 | 75.4 |

*lination, October,* 1889.

for any hour added to the base-line value gives the absolute westerly declination at that hour.

$= 2° 46' 26''$

| Day. | P. M. | | | | | | | | | | | | Mean. |
|---|---|---|---|---|---|---|---|---|---|---|---|---|---|
| | 1. | 2. | 3. | 4. | 5. | 6. | 7. | 8. | 9. | 10. | 11. | 12. | |
| 1 | 78.2 | 79.8 | 79.0 | 77.1 | 75.2 | 74.5 | 74.5 | 74.4 | 74.2 | 74.0 | 73.9 | 74.0 | 74.22 |
| 2 | 78.0 | 78.7 | 77.5 | 76.0 | 75.2 | 74.9 | 74.7 | 74.2 | 74.2 | 74.0 | 73.7 | 74.7 | 74.17 |
| 3 | 76.2 | 76.6 | 76.2 | 76.0 | 74.9 | 74.7 | 74.4 | 74.3 | 74.3 | 74.3 | 74.3 | 74.1 | 73.33 |
| 4 | 76.1 | 77.3 | 77.3 | 76.3 | 75.9 | 75.0 | 74.5 | 74.7 | 74.1 | 74.0 | 74.0 | 74.0 | 74.17 |
| 5 | 79.3 | 79.6 | 77.3 | 80.0 | 79.7 | 73.1 | 74.2 | 64.4 | 68.4 | 79.0 | 69.0 | 71.8 | 74.62 |
| 6 | 78.5 | 77.7 | 76.0 | 75.3 | 71.8 | 74.3 | 73.7 | 73.9 | 74.1 | 71.6 | 73.4 | 74.5 | 74.47 |
| 7 | 77.1 | 76.8 | 75.7 | 74.0 | 74.9 | 68.7 | 73.8 | 73.8 | 72.8 | 71.8 | 77.4 | 76.0 | 74.65 |
| 8 | 78.8 | 78.1 | 76.8 | 74.3 | 74.0 | 74.5 | 74.5 | 74.8 | 72.6 | 70.3 | 73.5 | 75.1 | 75.08 |
| 9 | 79.0 | 78.6 | 77.8 | 76.0 | 75.1 | 73.9 | 74.4 | 74.6 | 74.7 | 74.3 | 74.6 | 75.5 | 75.58 |
| 10 | 78.9 | 77.5 | 75.7 | 75.1 | 74.7 | 74.4 | 74.0 | 74.3 | 74.7 | 74.9 | 74.9 | 74.8 | 75.15 |
| 11 | 79.3 | 78.7 | 77.2 | 75.7 | 74.7 | 74.3 | 74.1 | 74.2 | 74.2 | 74.2 | 74.5 | 74.5 | 74.69 |
| 12 | 77.5 | 77.4 | 76.2 | 75.0 | 74.0 | 74.1 | 74.1 | 73.9 | 73.9 | 74.0 | 74.5 | 74.5 | 74.24 |
| 13 | 79.6 | 79.1 | 77.8 | 77.0 | 76.0 | 75.3 | 73.7 | 74.0 | 67.5 | 70.9 | 74.1 | 75.0 | 74.43 |
| 14 | 77.9 | 78.8 | 78.0 | 76.1 | 75.2 | 75.3 | 74.7 | 73.0 | 74.7 | 74.0 | 74.0 | 74.0 | 74.07 |
| 15 | 78.1 | 79.1 | 78.7 | 79.0 | 79.3 | 76.2 | 76.1 | 75.1 | 73.9 | 74.5 | 74.0 | 74.0 | 74.99 |
| 16 | 77.1 | 78.1 | 77.5 | 76.4 | 75.6 | 74.9 | 74.7 | 74.5 | 74.5 | 74.3 | 74.7 | 74.7 | 74.46 |
| 17 | 78.7 | 78.8 | 77.8 | 76.4 | 75.0 | 74.5 | 74.4 | 74.3 | 72.0 | 73.1 | 72.0 | 73.4 | 74.53 |
| 18 | 80.7 | 80.6 | 80.1 | 75.2 | 76.6 | 75.8 | 73.9 | 73.9 | 72.5 | 71.9 | 72.0 | 73.7 | 74.67 |
| 19 | 78.3 | 76.9 | 75.8 | 74.9 | 73.5 | 74.3 | 72.0 | 60.0 | 70.8 | 76.0 | 73.7 | 72.1 | 74.71 |
| 20 | 79.9 | 76.6 | 73.0 | 74.4 | 75.1 | 73.7 | 73.7 | 70.0 | 72.3 | 73.6 | 73.6 | 72.4 | 75.13 |
| 21 | 77.5 | 77.7 | 76.2 | 76.0 | 73.3 | 71.8 | 74.2 | 69.7 | 72.3 | 74.2 | 73.9 | 73.7 | 74.21 |
| 22 | 76.3 | 76.0 | 75.4 | 75.2 | 75.1 | 75.0 | 74.7 | 74.4 | 74.1 | 75.2 | 74.8 | 74.3 | 75.14 |
| 23 | 76.6 | 77.2 | 76.8 | 75.7 | 74.9 | 74.9 | 75.0 | 75.5 | 73.6 | 73.6 | 73.4 | 74.0 | 75.04 |
| 24 | 77.3 | 76.5 | 75.5 | 74.8 | 74.4 | 74.8 | 74.7 | 74.6 | 74.6 | 74.8 | 74.8 | 75.2 | 74.78 |
| 25 | 78.7 | 78.3 | 77.2 | 75.9 | 75.1 | 74.7 | 74.2 | 74.3 | 74.5 | 73.0 | 74.3 | 74.5 | 74.98 |
| 26 | 79.1 | 78.2 | 76.1 | 75.2 | 75.0 | 74.5 | 74.1 | 73.9 | 74.0 | 73.7 | 74.1 | 74.1 | 74.61 |
| 27 | 78.5 | 78.5 | 77.0 | 75.4 | 75.0 | 75.4 | 74.5 | 74.1 | 74.0 | 74.0 | 74.0 | 73.8 | 74.39 |
| 28 | 79.9 | 78.1 | 76.9 | 76.0 | 76.0 | 74.2 | 74.2 | 73.9 | 74.0 | 73.7 | 74.0 | 74.7 | 74.82 |
| 29 | 76.6 | 76.9 | 76.2 | 75.2 | 75.0 | 74.7 | 74.8 | 74.0 | 74.7 | 75.2 | 74.7 | 74.2 | 74.35 |
| 30 | 77.0 | 77.2 | 76.6 | 75.4 | 75.0 | 74.5 | 74.4 | 74.0 | 73.7 | 74.1 | 74.3 | 74.4 | 74.35 |
| 31 | 77.0 | 77.5 | 76.1 | 76.2 | 74.9 | 76.0 | 75.0 | 74.2 | 74.5 | 74.6 | 74.8 | 74.2 | 74.74 |

9230——3

MAGNETIC OBSERVATIONS AT THE U. S. NAVAL OBSERVATORY.

TABLE V—Continued—*Dec*

Ordinates, expressed in minutes of arc, taken from the daily declination traces. The ordinate

Base-line value

| Day. | A. M. | | | | | | | | | | | |
|------|-----|-----|-----|-----|-----|-----|-----|-----|-----|-----|-----|-----|
| | 1. | 2. | 3. | 4. | 5. | 6. | 7. | 8. | 9. | 10. | 11. | 12. |
| 1 | 74.0 | 70.3 | 69.3 | 68.1 | 72.1 | 74.0 | 75.2 | 77.5 | 80.0 | 78.0 | 89.3 | 79.1 |
| 2 | 72.2 | 80.5 | 76.0 | 76.0 | 75.3 | 74.0 | 76.1 | 73.0 | 75.3 | 76.0 | 75.2 | 75.1 |
| 3 | 75.8 | 75.0 | 74.0 | 79.5 | 73.1 | 74.0 | 74.0 | 78.0 | 75.0 | 74.3 | 75.3 | 76.0 |
| 4 | 76.0 | 74.4 | 77.0 | 75.1 | 73.4 | 73.1 | 72.8 | 73.4 | 73.8 | 75.3 | 76.9 | 77.5 |
| 5 | 76.9 | 76.1 | 76.0 | 74.1 | 73.3 | 75.8 | 77.8 | 74.2 | 75.3 | 77.4 | 79.0 | 78.6 |
| 6 | 75.2 | 76.3 | 76.0 | 74.3 | 74.0 | 74.0 | 73.0 | 72.7 | 72.3 | 74.2 | 76.2 | 77.9 |
| 7 | 75.6 | 75.4 | 75.4 | 75.2 | 75.2 | 75.0 | 74.8 | 73.2 | 74.2 | 75.9 | 78.1 | 79.7 |
| 8 | 74.4 | 73.7 | 74.3 | 74.3 | 73.0 | 72.2 | 71.8 | 71.0 | 71.0 | 72.4 | 74.6 | 76.5 |
| 9 | 74.2 | 74.0 | 74.0 | 73.6 | 73.2 | 73.0 | 72.7 | 72.2 | 72.2 | 73.7 | 75.7 | 77.7 |
| 10 | 73.5 | 73.2 | 73.9 | 73.2 | 72.4 | 72.5 | 72.0 | 71.3 | 71.2 | 72.7 | 74.3 | 75.7 |
| 11 | 74.0 | 74.0 | 75.4 | 73.1 | 72.9 | 72.1 | 71.7 | 71.0 | 71.6 | 73.5 | 75.4 | 76.7 |
| 12 | 73.8 | 74.0 | 72.7 | 73.1 | 72.4 | 72.3 | 72.0 | 71.4 | 71.4 | 72.7 | 74.0 | 75.4 |
| 13 | 73.6 | 73.5 | 73.5 | 73.5 | 73.4 | 72.8 | 72.8 | 72.1 | 72.0 | 72.1 | 73.0 | 74.1 |
| 14 | 73.1 | 73.4 | 73.3 | 73.1 | 73.3 | 72.9 | 72.4 | 71.9 | 71.5 | 71.7 | 72.7 | 73.9 |
| 15 | 73.4 | 73.5 | 73.7 | 73.6 | 73.0 | 72.9 | 72.8 | 72.1 | 70.6 | 71.7 | 73.2 | 75.3 |
| 16 | 74.3 | 73.0 | 72.8 | 71.0 | 72.4 | 72.0 | 71.9 | 71.0 | 71.9 | 73.4 | 74.8 | 75.5 |
| 17 | 73.5 | 74.5 | 73.1 | 79.8 | 70.9 | 74.2 | 72.1 | 73.3 | 72.1 | 73.6 | 75.3 | 76.6 |
| 18 | 79.8 | 72.8 | 76.2 | 72.3 | 72.6 | 72.4 | 72.4 | 72.0 | 72.3 | 73.0 | 75.5 | 75.9 |
| 19 | 74.9 | 74.5 | 73.9 | 73.2 | 73.2 | 73.0 | 72.4 | 72.2 | 72.0 | 72.9 | 73.5 | 74.3 |
| 20 | 73.6 | 73.5 | 73.6 | 73.5 | 73.2 | 72.8 | 73.1 | 71.9 | 71.3 | 72.2 | 73.5 | 75.2 |
| 21 | 73.9 | 73.7 | 73.2 | 73.3 | 72.3 | 72.0 | 72.5 | 71.1 | 70.4 | 71.4 | 73.0 | 74.6 |
| 22 | 73.3 | 73.4 | 77.9 | 72.2 | 71.9 | 71.5 | 73.7 | 71.0 | 70.3 | 71.4 | 73.2 | 74.8 |
| 23 | 73.4 | 73.9 | 73.1 | 73.1 | 73.1 | 72.5 | 72.2 | 71.5 | 71.2 | 71.9 | 73.4 | 75.5 |
| 24 | 72.6 | 73.6 | 72.9 | 72.5 | 72.2 | 72.0 | 71.6 | 71.2 | 71.0 | 71.5 | 73.7 | 75.4 |
| 25 | 73.1 | 74.7 | 75.7 | 71.9 | 72.1 | 72.6 | 72.8 | 71.1 | 70.0 | 71.8 | 73.8 | 76.1 |
| 26 | 73.0 | 74.2 | 72.7 | 72.5 | 71.8 | 75.0 | 78.9 | 75.6 | 73.4 | 74.6 | 76.1 | 77.0 |
| 27 | 68.0 | 70.7 | 72.4 | 73.6 | 73.6 | 72.8 | 72.0 | 72.0 | 70.9 | 70.2 | 73.0 | 74.6 |
| 28 | 73.3 | 73.4 | 73.3 | 73.2 | 73.1 | 73.3 | 72.2 | 70.6 | 69.0 | 72.3 | 74.8 | 74.0 |
| 29 | 72.0 | 73.1 | 74.6 | 75.1 | 74.9 | 74.1 | 73.6 | 73.9 | 72.0 | 72.1 | 76.3 | 77.3 |
| 30 | 73.7 | 75.2 | 75.8 | 72.6 | 73.8 | 79.8 | 76.9 | 72.1 | 70.0 | 70.7 | 73.1 | 74.1 |

*lination, November,* 1889.

for any hour added to the base-line value gives the absolute westerly declination at that hour.

$= 2° 47' 23''$

| Day. | P. M. | | | | | | | | | | | | Mean. |
|---|---|---|---|---|---|---|---|---|---|---|---|---|---|
| | 1. | 2. | 3. | 4. | 5. | 6. | 7. | 8. | 9. | 10. | 11. | 12. | |
| 1 | 79.6 | 77.8 | 79.0 | 76.6 | 75.5 | 70.4 | 58.8 | 71.9 | 70.8 | 68.1 | 74.0 | 74.9 | 74.35 |
| 2 | 77.3 | 77.4 | 76.2 | 75.0 | 73.5 | 73.0 | 74.7 | 73.5 | 69.1 | 73.3 | 73.0 | 78.0 | 74.95 |
| 3 | 76.7 | 75.7 | 76.1 | 75.0 | 73.5 | 74.0 | 74.0 | 73.4 | 70.7 | 70.0 | 75.2 | 74.8 | 74.71 |
| 4 | 77.4 | 77.0 | 76.0 | 75.1 | 74.9 | 74.7 | 75.0 | 74.7 | 73.4 | 73.2 | 74.0 | 76.1 | 75.01 |
| 5 | 77.8 | 77.7 | 76.3 | 75.5 | 75.1 | 74.9 | 74.7 | 63.0 | 72.0 | 73.4 | 74.7 | 75.2 | 75.20 |
| 6 | 78.1 | 77.1 | 76.0 | 75.0 | 74.7 | 74.5 | 74.5 | 74.4 | 74.7 | 74.5 | 74.0 | 75.0 | 74.94 |
| 7 | 78.2 | 77.1 | 75.1 | 73.7 | 73.6 | 73.1 | 73.2 | 73.0 | 73.3 | 73.3 | 73.1 | 73.8 | 74.93 |
| 8 | 76.6 | 76.0 | 74.5 | 74.0 | 73.6 | 73.6 | 73.3 | 73.1 | 73.2 | 73.2 | 74.0 | 74.2 | 73.69 |
| 9 | 78.0 | 77.3 | 76.0 | 74.7 | 73.8 | 74.3 | 75.8 | 72.4 | 72.2 | 72.6 | 72.0 | 75.6 | 74.20 |
| 10 | 76.5 | 76.1 | 76.0 | 74.9 | 74.2 | 73.3 | 72.4 | 73.0 | 72.7 | 72.9 | 73.5 | 73.2 | 73.53 |
| 11 | 77.2 | 76.6 | 76.4 | 75.4 | 74.1 | 73.1 | 73.1 | 72.7 | 73.4 | 73.0 | 73.1 | 73.0 | 73.87 |
| 12 | 76.7 | 76.1 | 76.2 | 75.6 | 74.3 | 73.9 | 73.1 | 73.3 | 73.1 | 73.3 | 73.1 | 73.2 | 73.63 |
| 13 | 75.1 | 75.4 | 74.7 | 74.1 | 73.8 | 73.1 | 72.9 | 72.7 | 72.3 | 73.0 | 73.0 | 73.1 | 73.32 |
| 14 | 75.0 | 75.8 | 75.5 | 74.3 | 73.1 | 72.8 | 72.3 | 72.4 | 72.9 | 73.0 | 72.8 | 73.2 | 73.18 |
| 15 | 76.3 | 76.7 | 76.9 | 76.0 | 75.0 | 73.2 | 73.6 | 72.9 | 73.1 | 72.8 | 73.3 | 73.1 | 73.70 |
| 16 | 75.5 | 75.6 | 75.0 | 75.1 | 74.1 | 73.7 | 73.3 | 73.0 | 72.0 | 71.9 | 73.0 | 73.5 | 73.32 |
| 17 | 78.8 | 78.2 | 78.0 | 77.4 | 75.0 | 74.0 | 72.7 | 71.1 | 68.7 | 68.9 | 72.8 | 73.6 | 74.09 |
| 18 | 76.5 | 75.4 | 74.8 | 74.4 | 73.9 | 73.2 | 72.9 | 72.8 | 72.8 | 73.0 | 71.1 | 75.0 | 73.88 |
| 19 | 75.0 | 74.7 | 74.0 | 73.1 | 73.1 | 73.0 | 73.0 | 72.8 | 73.0 | 73.0 | 73.2 | 73.4 | 73.39 |
| 20 | 75.7 | 74.9 | 73.9 | 73.3 | 73.0 | 72.8 | 72.7 | 72.6 | 73.0 | 73.0 | 71.9 | 72.5 | 73.20 |
| 21 | 75.8 | 75.3 | 74.2 | 73.2 | 72.9 | 72.1 | 71.6 | 71.8 | 71.6 | 72.7 | 73.2 | 72.9 | 72.86 |
| 22 | 75.5 | 75.1 | 74.4 | 73.3 | 72.9 | 72.5 | 72.9 | 72.5 | 72.2 | 71.1 | 72.0 | 72.3 | 72.97 |
| 23 | 77.0 | 76.0 | 74.9 | 74.0 | 73.0 | 72.8 | 72.6 | 72.3 | 72.4 | 72.5 | 72.5 | 73.4 | 73.26 |
| 24 | 76.9 | 75.8 | 75.5 | 77.4 | 73.0 | 73.1 | 72.3 | 73.1 | 69.0 | 73.1 | 72.2 | 73.0 | 73.13 |
| 25 | 77.0 | 74.3 | 74.0 | 73.1 | 73.0 | 73.2 | 73.4 | 69.0 | 72.4 | 72.0 | 72.1 | 72.1 | 72.97 |
| 26 | 78.9 | 77.0 | 79.4 | 78.5 | 77.1 | 79.6 | 77.7 | 73.1 | 70.3 | 69.1 | 70.0 | 73.0 | 74.94 |
| 27 | 77.4 | 79.5 | 80.5 | 80.2 | 79.5 | 80.0 | 79.6 | 73.9 | 72.1 | 71.2 | 71.9 | 73.0 | 74.28 |
| 28 | 78.0 | 78.5 | 76.0 | 74.5 | 73.4 | 71.0 | 72.0 | 70.8 | 69.4 | 70.0 | 71.0 | 73.0 | 72.92 |
| 29 | 76.4 | 76.5 | 75.1 | 74.5 | 73.5 | 71.4 | 71.7 | 72.3 | 72.1 | 71.6 | 73.4 | 71.8 | 73.72 |
| 30 | 74.2 | 75.9 | 75.2 | 74.2 | 73.4 | 73.5 | 73.3 | 72.2 | 71.3 | 72.0 | 73.1 | 73.6 | 73.74 |

TABLE V—Continued—*Dec*

Ordinates, expressed in minutes of arc, taken from the daily declination traces.  The ordinate

Base-line value

| Day. | A. M. | | | | | | | | | | | |
|---|---|---|---|---|---|---|---|---|---|---|---|---|
| | 1. | 2. | 3. | 4. | 5. | 6. | 7. | 8. | 9. | 10. | 11. | 12. |
| | ′ | ′ | ′ | ′ | ′ | ′ | ′ | ′ | ′ | ′ | ′ | ′ |
| 1 | 73.8 | 73.6 | 75.0 | 73.6 | 72.3 | 74.3 | 73.1 | 72.1 | 71.6 | 70.9 | 71.5 | 73.8 |
| 2 | 74.0 | 74.0 | 74.4 | 73.7 | 72.8 | 73.7 | 72.8 | 72.5 | 72.1 | 72.0 | 73.2 | 75.1 |
| 3 | 73.8 | 74.1 | 73.9 | 72.7 | 73.2 | 73.3 | 72.9 | 73.5 | 71.9 | 71.9 | 73.2 | 75.2 |
| 4 | 73.4 | 73.1 | 73.2 | 73.9 | 73.0 | 72.9 | 72.9 | 72.4 | 72.0 | 72.1 | 72.6 | 73.6 |
| 5 | 73.6 | 73.3 | 73.1 | 73.0 | 73.3 | 73.0 | 73.0 | 72.7 | 72.1 | 72.0 | 72.4 | 74.0 |
| 6 | 72.4 | 73.4 | 73.0 | 72.9 | 72.3 | 72.9 | 71.3 | 71.0 | 68.6 | 70.0 | 74.9 | 76.9 |
| 7 | 74.0 | 73.6 | 72.1 | 72.7 | 73.0 | 71.2 | 72.4 | 71.0 | 71.3 | 72.1 | 74.2 | 76.6 |
| 8 | 74.3 | 74.1 | 73.6 | 72.6 | 72.6 | 72.0 | 72.0 | 70.6 | 70.9 | 72.4 | 74.0 | 75.0 |
| 9 | 73.9 | 74.0 | 73.4 | 71.9 | 73.0 | 72.3 | 71.7 | 70.9 | 70.4 | 72.0 | 73.7 | 75.9 |
| 10 | 72.8 | 73.7 | 73.6 | 72.7 | 73.2 | 73.8 | 71.7 | 71.3 | 71.1 | 71.1 | 72.7 | 74.8 |
| 11 | 73.4 | 73.4 | 73.5 | 73.2 | 73.0 | 72.6 | 72.5 | 72.0 | 71.5 | 71.9 | 73.1 | 74.9 |
| 12 | 73.2 | 73.4 | 73.5 | 73.4 | 73.2 | 72.9 | 72.5 | 72.0 | 71.6 | 71.3 | 72.0 | 73.6 |
| 13 | 73.1 | 73.1 | 73.4 | 73.7 | 71.3 | 72.7 | 72.7 | 72.2 | 70.6 | 71.5 | 72.0 | 73.3 |
| 14 | 77.5 | 71.1 | 71.4 | 72.5 | 73.1 | 73.8 | 73.1 | 72.5 | 73.1 | 73.0 | 72.5 | 73.3 |
| 15 | 73.4 | 74.0 | 74.0 | 73.1 | 73.4 | 74.0 | 73.7 | 72.3 | 71.4 | 70.9 | 72.0 | 74.1 |
| 16 | 73.5 | 73.5 | 73.7 | 73.6 | 73.4 | 73.4 | 72.4 | 72.0 | 70.6 | 70.3 | 72.8 | 74.8 |
| 17 | 74.0 | 73.7 | 74.1 | 73.4 | 75.6 | 74.8 | 72.4 | 72.0 | 71.3 | 72.6 | 73.5 | 74.5 |
| 18 | 73.5 | 73.5 | 73.5 | 73.5 | 74.1 | 73.4 | 72.6 | 72.0 | 71.6 | 71.0 | 72.0 | 73.5 |
| 19 | 72.7 | 72.1 | 73.0 | 72.0 | 72.9 | 73.4 | 72.2 | 71.8 | 70.0 | 69.4 | 72.3 | 73.3 |
| 20 | 73.4 | 72.4 | 72.5 | 73.9 | 72.2 | 71.4 | 72.3 | 75.0 | 72.1 | 69.6 | 70.8 | 73.1 |
| 21 | 72.6 | 73.1 | 72.5 | 71.8 | 73.5 | 72.8 | 72.5 | 71.3 | 70.1 | 70.3 | 71.7 | 73.6 |
| 22 | 73.4 | 72.3 | 73.1 | 73.2 | 72.5 | 72.9 | 74.0 | 72.0 | 71.0 | 72.0 | 73.8 | 76.0 |
| 23 | 72.3 | 73.8 | 72.9 | 72.7 | 72.7 | 74.0 | 73.5 | 72.0 | 69.5 | 70.0 | 71.2 | 72.9 |
| 24 | 73.3 | 73.2 | 73.0 | 73.0 | 73.1 | 72.8 | 72.2 | 70.9 | 69.0 | 70.7 | 73.5 | 75.4 |
| 25 | 73.9 | 73.5 | 73.1 | 73.6 | 73.6 | 72.1 | 72.0 | 70.7 | 69.5 | 70.0 | 72.6 | 75.1 |
| 26 | 73.0 | 73.5 | 73.4 | 73.3 | 74.0 | 72.9 | 72.9 | 70.9 | 70.0 | 69.7 | 71.7 | 75.0 |
| 27 | 66.4 | 71.7 | 73.0 | 71.3 | 74.3 | 73.1 | 72.9 | 70.9 | 68.3 | 68.4 | 71.8 | 74.7 |
| 28 | 73.9 | 74.1 | 74.0 | 74.1 | 74.0 | 73.4 | 73.6 | 72.0 | 69.1 | 70.0 | 75.2 | 76.9 |
| 29 | 73.2 | 77.1 | 74.4 | 73.3 | 74.2 | 74.5 | 73.0 | 71.8 | 70.5 | 70.4 | 72.3 | 75.2 |
| 30 | 73.4 | 74.0 | 74.0 | 73.4 | 73.4 | 74.5 | .74.4 | 72.2 | 72.7 | 72.7 | 74.0 | 75.5 |
| 31 | 71.0 | 71.5 | 69.9 | 71.1 | 71.2 | 71.3 | 71.8 | 71.0 | 70.2 | 69.8 | . . . | . . . |

*lination, December,* 1889.

for any hour added to the base-line value gives the absolute westerly declination at that hour.

$= 2°\ 48'\ 52''$

| Day. | P. M. | | | | | | | | | | | | Mean. |
|---|---|---|---|---|---|---|---|---|---|---|---|---|---|
| | 1. | 2. | 3. | 4. | 5. | 6. | 7. | 8. | 9. | 10. | 11. | 12. | |
| 1 | 74.8 | 75.3 | 74.8 | 75.0 | 73.9 | 73.0 | 73.2 | 72.8 | 72.5 | 73.0 | 73.4 | 73.1 | 73.35 |
| 2 | 75.8 | 75.4 | 75.0 | 74.0 | 73.2 | 73.0 | 71.4 | 72.8 | 72.6 | 73.0 | 73.0 | 73.5 | 73.46 |
| 3 | 74.4 | 73.8 | 74.0 | 74.0 | 71.9 | 72.0 | 72.6 | 72.5 | 72.6 | 73.0 | 73.2 | 73.4 | 73.21 |
| 4 | 74.1 | 74.8 | 74.4 | 74.0 | 73.2 | 73.1 | 73.0 | 72.7 | 72.5 | 73.2 | 73.3 | 73.4 | 73.20 |
| 5 | 74.9 | 74.7 | 74.3 | 74.0 | 73.9 | 73.1 | 72.7 | 72.2 | 72.2 | 72.7 | 72.6 | 73.5 | 73.18 |
| 6 | 77.3 | 79.0 | 74.1 | 73.4 | 73.4 | 71.6 | 72.0 | 72.0 | 73.1 | 71.4 | 72.8 | 74.0 | 73.07 |
| 7 | 77.5 | 74.6 | 75.0 | 73.0 | 74.0 | 72.6 | 73.0 | 71.2 | 73.4 | 71.9 | 72.5 | 72.5 | 73.14 |
| 8 | 75.9 | 75.6 | 74.7 | 73.4 | 73.0 | 72.4 | 72.4 | 72.4 | 71.0 | 72.2 | 72.8 | 73.0 | 73.04 |
| 9 | 76.9 | 76.7 | 75.7 | 74.0 | 73.4 | 72.5 | 72.1 | 72.1 | 72.1 | 72.3 | 72.7 | 72.9 | 73.19 |
| 10 | 76.2 | 75.9 | 74.0 | 73.0 | 72.7 | 72.9 | 72.8 | 72.1 | 71.6 | 73.2 | 73.0 | 73.2 | 73.05 |
| 11 | 75.4 | 75.2 | 75.0 | 74.1 | 73.0 | 72.4 | 72.4 | 72.3 | 72.0 | 72.5 | 72.8 | 72.6 | 73.13 |
| 12 | 74.5 | 74.4 | 74.0 | 73.9 | 73.4 | 72.7 | 72.8 | 72.2 | 72.5 | 72.1 | 72.3 | 72.6 | 72.92 |
| 13 | 74.2 | 74.5 | 74.5 | 74.4 | 74.4 | 73.3 | 72.5 | 72.0 | 72.5 | 72.5 | 71.0 | 69.8 | 72.69 |
| 14 | 73.0 | 73.5 | 73.4 | 73.7 | 73.2 | 73.1 | 72.9 | 72.4 | 72.4 | 72.4 | 72.4 | 73.2 | 73.02 |
| 15 | 74.6 | 74.3 | 74.1 | 73.9 | 73.4 | 73.0 | 72.9 | 72.4 | 72.9 | 72.3 | 73.0 | 73.2 | 73.18 |
| 16 | 75.2 | 74.9 | 74.1 | 73.3 | 73.1 | 73.9 | 71.0 | 71.6 | 72.2 | 72.5 | 72.1 | 73.1 | 72.96 |
| 17 | 75.0 | 75.1 | 75.0 | 74.0 | 73.0 | 72.0 | 72.8 | 72.5 | 72.7 | 72.7 | 72.8 | 72.9 | 73.42 |
| 18 | 75.0 | 75.1 | 74.8 | 74.0 | 73.1 | 73.0 | 72.5 | 72.8 | 72.5 | 71.7 | 72.0 | 72.7 | 73.06 |
| 19 | 73.1 | 74.6 | 74.8 | 74.8 | 73.1 | 73.3 | .72.5 | 72.5 | 72.7 | 72.1 | 74.0 | 72.0 | 72.69 |
| 20 | 74.1 | 74.2 | 74.4 | 74.2 | 73.4 | 73.0 | 73.3 | 72.2 | 72.4 | 72.8 | 72.8 | 73.3 | 72.87 |
| 21 | 74.7 | 75.1 | 75.1 | 75.2 | 73.8 | 73.6 | 72.9 | 72.2 | 72.1 | 69.0 | 72.2 | 74.8 | 72.77 |
| 22 | 74.6 | 74.6 | 77.1 | 75.6 | 74.0 | 74.0 | 73.7 | 72.4 | 72.0 | 71.5 | 72.5 | 71.0 | 73.30 |
| 23 | 74.1 | 74.8 | 74.7 | 74.2 | 73.3 | 72.9 | 72.8 | 72.4 | 72.4 | 71.6 | 72.1 | 73.0 | 72.74 |
| 24 | 75.9 | 74.6 | 74.4 | 73.5 | 73.4 | 73.0 | 72.9 | 71.6 | 71.8 | 72.9 | 72.3 | 73.1 | 72.90 |
| 25 | 75.6 | 74.9 | 74.0 | 73.0 | 72.5 | 72.4 | 72.4 | 72.7 | 72.7 | 72.7 | 73.0 | 73.0 | 72.86 |
| 26 | 76.5 | 76.9 | 75.0 | 74.5 | 73.2 | 72.0 | 71.2 | 60.2 | 69.7 | 69.0 | 69.3 | 68.1 | 72.42 |
| 27 | 75.4 | 75.8 | 75.0 | 73.6 | 73.0 | 72.6 | 72.4 | 72.3 | 72.5 | 72.8 | 73.1 | 73.8 | 72.46 |
| 28 | 76.9 | 76.0 | 75.1 | 73.4 | 73.0 | 73.5 | 72.9 | 72.2 | 72.5 | 70.9 | 71.5 | 72.7 | 73.37 |
| 29 | 76.4 | 77.0 | 76.6 | 76.0 | 72.8 | 73.0 | 72.8 | 72.3 | 69.0 | 68.3 | 73.1 | 73.1 | 73.35 |
| 30 | 74.5 | 75.1 | 74.2 | 72.1 | 71.3 | 71.0 | 70.5 | 70.4 | 70.4 | 70.5 | 70.6 | 71.0 | 72.74 |
| 31 | . . . | . . . | . . . | . . . | 72.1 | 71.6 | 71.1 | 70.2 | 70.6 | 70.6 | 70.8 | 71.2 | 70.96 |

TABLE VI.—*Horizontal*

The figures given in the table are millionths of a dyne, which, added

. 198000 +

| Day. | | | | | | A. M. | | | | | | |
|---|---|---|---|---|---|---|---|---|---|---|---|---|
| | 1. | 2. | 3. | 4. | 5. | 6. | 7. | 8. | 9. | 10. | 11. | 12. |
| 1 | ... | ... | ... | ... | ... | ... | ... | ... | ... | ... | ... | ... |
| 2 | ... | ... | ... | ... | ... | ... | ... | ... | ... | ... | ... | ... |
| 3 | ... | ... | ... | ... | ... | ... | ... | ... | ... | ... | ... | ... |
| 4 | ... | ... | ... | ... | ... | ... | ... | ... | ... | ... | ... | ... |
| 5 | ... | ... | ... | ... | ... | ... | ... | ... | ... | ... | ... | ... |
| 6 | ... | ... | ... | ... | ... | ... | ... | ... | ... | ... | ... | ... |
| 7 | ... | ... | ... | ... | ... | ... | ... | ... | ... | ... | ... | ... |
| 8 | ... | ... | ... | ... | ... | ... | ... | ... | ... | ... | ... | ... |
| 9 | ... | ... | ... | ... | ... | ... | ... | ... | ... | ... | ... | ... |
| 10 | ... | ... | ... | ... | ... | ... | ... | ... | ... | ... | ... | ... |
| 11 | ... | ... | ... | ... | ... | ... | ... | ... | ... | ... | ... | ... |
| 12 | ... | ... | ... | ... | ... | ... | ... | ... | ... | ... | ... | ... |
| 13 | ... | ... | ... | ... | ... | ... | ... | ... | ... | ... | ... | ... |
| 14 | ... | ... | ... | ... | ... | ... | ... | ... | ... | ... | ... | ... |
| 15 | ... | ... | ... | ... | ... | ... | ... | ... | ... | ... | ... | ... |
| 16 | ... | ... | ... | ... | ... | ... | ... | ... | ... | ... | ... | ... |
| 17 | ... | ... | ... | ... | ... | ... | ... | ... | ... | ... | ... | ... |
| 18 | ... | ... | ... | ... | ... | ... | ... | ... | ... | ... | ... | ... |
| 19 | ... | ... | ... | ... | ... | ... | ... | ... | ... | ... | ... | ... |
| 20 | ... | ... | ... | ... | ... | ... | ... | ... | ... | ... | ... | ... |
| 21 | ... | ... | ... | ... | ... | ... | ... | ... | ... | ... | ... | ... |
| 22 | ... | ... | ... | ... | ... | ... | ... | ... | ... | ... | ... | ... |
| 23 | ... | ... | ... | ... | ... | ... | ... | ... | ... | ... | ... | ... |
| 24 | ... | ... | ... | ... | ... | ... | ... | ... | ... | ... | ... | ... |
| 25 | ... | ... | ... | ... | ... | ... | ... | ... | ... | ... | ... | ... |
| 26 | ... | ... | ... | ... | ... | ... | ... | ... | ... | ... | ... | ... |
| 27 | ... | ... | ... | ... | ... | ... | ... | ... | ... | ... | ... | ... |
| 28 | 934 | 948 | 897 | 934 | 920 | 972 | 981 | 1000 | 1061 | 1071 | 991 | 864 |
| 29 | 974 | 950 | 950 | 1044 | 1007 | 988 | 974 | 974 | 1040 | 1012 | 889 | 800 |
| 30 | 942 | 933 | 919 | 975 | 989 | 1027 | 1003 | 1045 | 1050 | 1074 | 1003 | 984 |
| 31 | 892 | 906 | 963 | 921 | 916 | 930 | 944 | 954 | 925 | 921 | 869 | 864 |
| Mean . | 935 | 934 | 932 | 968 | 958 | 979 | 975 | 993 | 1019 | 1019 | 938 | 878 |

*force*, **January,** 1889.

to .198000 dyne, give the absolute horizontal force in C. G. S. units.

. 198000 +

| Day. | P. M. | | | | | | | | | | | | Mean. |
| | 1. | 2. | 3. | 4. | 5. | 6. | 7. | 8. | 9. | 10. | 11. | 12. | |
|---|---|---|---|---|---|---|---|---|---|---|---|---|---|
| 1 | . . . | . . . | . . . | . . . | . . . | . . . | . . . | . . . | . . . | . . . | . . . | . . . | . . . |
| 2 | . . . | . . . | . . . | . . . | . . . | . . . | . . . | . . . | . . . | . . . | . . . | . . . | . . . |
| 3 | . . . | . . . | . . . | . . . | . . . | . . . | . . . | . . . | . . . | . . . | . . . | . . . | . . . |
| 4 | . . . | . . . | . . . | . . . | . . . | . . . | . . . | . . . | . . . | . . . | . . . | . . . | . . . |
| 5 | . . . | . . . | . . . | . . . | . . . | . . . | . . . | . . . | . . . | . . . | . . . | . . . | . . . |
| 6 | . . . | . . . | . . . | . . . | . . . | . . . | . . . | . . . | . . . | . . . | . . . | . . . | . . . |
| 7 | . . . | . . . | . . . | . . . | . . . | . . . | . . . | . . . | . . . | . . . | . . . | . . . | . . . |
| 8 | . . . | . . . | . . . | . . . | . . . | . . . | . . . | . . . | . . . | . . . | . . . | . . . | . . . |
| 9 | . . . | . . . | . . . | . . . | . . . | . . . | . . . | . . . | . . . | . . . | . . . | . . . | . . . |
| 10 | . . . | . . . | . . . | . . . | . . . | . . . | . . . | . . . | . . . | . . . | . . . | . . . | . . . |
| 11 | . . . | . . . | . . . | . . . | . . . | . . . | . . . | . . . | . . . | . . . | . . . | . . . | . . . |
| 12 | . . . | . . . | . . . | . . . | . . . | . . . | . . . | . . . | . . . | . . . | . . . | . . . | . . . |
| 13 | . . . | . . . | . . . | . . . | . . . | . . . | . . . | . . . | . . . | . . . | . . . | . . . | . . . |
| 14 | . . . | . . . | . . . | . . . | . . . | . . . | . . . | . . . | . . . | . . . | . . . | . . . | . . . |
| 15 | . . . | . . . | . . . | . . . | . . . | . . . | . . . | . . . | . . . | . . . | . . . | . . . | . . . |
| 16 | . . . | . . . | . . . | . . . | . . . | . . . | . . . | . . . | . . . | . . . | . . . | . . . | . . . |
| 17 | . . . | . . . | . . . | . . . | . . . | . . . | . . . | . . . | . . . | . . . | . . . | . . . | . . . |
| 18 | . . . | . . . | . . . | . . . | . . . | . . . | . . . | . . . | . . . | . . . | . . . | . . . | . . . |
| 19 | . . . | . . . | . . . | . . . | . . . | . . . | . . . | . . . | . . . | . . . | . . . | . . . | . . . |
| 20 | . . . | . . . | . . . | . . . | . . . | . . . | . . . | . . . | . . . | . . . | . . . | . . . | . . . |
| 21 | . . . | . . . | . . . | . . . | . . . | . . . | . . . | . . . | . . . | . . . | . . . | . . . | . . . |
| 22 | . . . | . . . | . . . | . . . | . . . | . . . | . . . | . . . | . . . | . . . | . . . | . . . | . . . |
| 23 | . . . | . . . | . . . | . . . | . . . | . . . | . . . | . . . | . . . | . . . | . . . | . . . | . . . |
| 24 | . . . | . . . | . . . | . . . | . . . | . . . | . . . | . . . | . . . | . . . | . . . | . . . | . . . |
| 25 | . . . | . . . | . . . | . . . | . . . | . . . | . . . | . . . | . . . | . . . | . . . | . . . | . . . |
| 26 | . . . | . . . | . . . | . . . | . . . | . . . | . . . | . . . | . . . | . . . | . . . | . . . | . . . |
| 27 | . . . | . . . | . . . | . . . | . . . | . . . | . . . | . . . | . . . | . . . | . . . | . . . | . . . |
| 28 | 794 | 776 | 818 | 860 | 926 | 949 | 954 | 954 | 968 | 964 | 912 | 926 | 932 |
| 29 | 872 | 890 | 862 | 886 | 909 | 956 | 980 | 966 | 975 | 989 | 975 | 951 | 951 |
| 30 | 971 | 905 | 854 | 760 | 891 | 849 | 882 | 915 | 868 | 873 | 873 | 849 | 935 |
| 31 | 903 | 992 | 1011 | 969 | 936 | 954 | 950 | 889 | 832 | 893 | 799 | 898 | 918 |
| Mean | 885 | 891 | 886 | 869 | 915 | 927 | 942 | 931 | 911 | 930 | 890 | 906 | 934 |

TABLE VI—Continued—*Hori*

The figures given in the table are millionths of a dyne, which, added

.198000 +

| Day. | A. M. | | | | | | | | | | | |
|---|---|---|---|---|---|---|---|---|---|---|---|---|
| | 1. | 2. | 3. | 4. | 5. | 6. | 7. | 8. | 9. | 10. | 11. | 12. |
| 1 | 906 | 906 | 943 | 920 | 939 | 981 | 976 | 929 | 901 | 835 | 751 | 708 |
| 2 | 916 | 911 | 907 | 916 | 911 | 926 | 944 | 958 | 954 | 888 | 780 | 728 |
| 3 | 895 | 946 | 956 | 1022 | 1022 | 1022 | 1036 | 1012 | 1026 | 984 | 848 | 763 |
| 4 | 859 | 854 | 901 | 859 | 873 | 878 | 897 | 897 | 817 | 742 | 638 | 633 |
| 5 | 913 | 903 | 903 | 932 | 946 | 969 | 964 | 932 | 903 | 823 | 720 | 682 |
| 6 | 872 | 924 | 929 | 948 | 966 | 981 | 1013 | 1013 | 976 | 863 | 807 | 811 |
| 7 | 851 | 832 | 851 | 884 | 860 | 931 | 940 | 978 | 992 | 743 | 889 | 842 |
| 8 | 811 | 778 | 820 | 768 | 717 | 862 | 909 | 787 | 844 | 712 | 665 | 646 |
| 9 | 817 | 827 | 831 | 850 | 846 | 883 | 916 | 925 | 925 | 850 | 770 | 629 |
| 10 | 829 | 843 | 843 | 871 | 880 | 909 | 951 | 909 | 899 | 848 | 768 | 707 |
| 11 | 864 | 878 | 892 | 901 | 915 | 934 | 948 | 953 | 995 | 897 | 807 | 793 |
| 12 | 842 | 842 | 875 | 889 | 913 | 917 | 903 | 847 | 772 | 725 | 687 | 734 |
| 13 | 882 | 886 | 886 | 910 | 933 | 943 | 933 | 938 | 910 | 868 | 854 | 807 |
| 14 | 907 | 841 | 893 | 888 | 804 | 860 | 865 | 893 | 888 | 757 | 790 | 827 |
| 15 | 824 | 800 | 880 | 885 | 861 | 866 | 880 | 922 | 871 | 861 | 753 | 748 |
| 16 | 721 | 754 | 759 | 815 | 835 | 873 | 873 | 835 | 784 | 549 | 648 | 704 |
| 17 | 783 | 769 | 830 | 844 | 802 | 831 | 929 | 591 | 968 | 930 | 860 | 747 |
| 18 | 634 | 704 | 681 | 690 | 672 | 686 | 738 | 714 | 598 | 701 | 673 | 659 |
| 19 | 790 | 795 | 785 | 790 | 777 | 800 | 876 | 876 | 881 | 844 | 801 | 717 |
| 20 | 720 | 682 | 776 | 767 | 768 | 777 | 787 | 829 | 741 | 708 | 637 | 487 |
| 21 | 777 | 749 | 777 | 796 | 811 | 821 | 830 | 825 | 779 | 770 | 742 | 714 |
| 22 | 802 | 821 | 821 | 821 | 836 | 803 | 831 | 803 | 776 | 823 | 729 | 701 |
| 23 | 765 | 746 | 779 | 803 | 832 | 832 | 865 | 883 | 795 | 696 | 696 | 682 |
| 24 | 813 | 809 | 827 | 837 | 838 | 852 | 875 | 908 | 867 | 839 | 825 | 792 |
| 25 | 866 | 890 | 899 | 909 | 938 | 947 | 872 | 975 | 972 | 887 | 882 | 840 |
| 26 | 881 | 876 | 890 | 904 | 905 | 915 | 896 | 886 | 897 | 901 | 854 | 784 |
| 27 | 793 | 802 | 844 | 835 | 911 | 859 | 883 | 892 | 898 | 884 | 922 | 893 |
| 28 | 832 | 803 | 803 | 813 | 828 | 818 | 809 | 757 | 725 | 716 | 693 | 716 |
| Mean . | 827 | 828 | 849 | 860 | 862 | 881 | 898 | 881 | 870 | 809 | 768 | 732 |

*zontal force, February,* 1889.

to .198000 dyne, give the absolute horizontal force in C. G. S. units.

.198000 +

| Day. | P. M. | | | | | | | | | | | | Mean. |
|---|---|---|---|---|---|---|---|---|---|---|---|---|---|
| | 1. | 2. | 3. | 4. | 5. | 6. | 7. | 8. | 9. | 10. | 11. | 12. | |
| 1 | 846 | 888 | 921 | 954 | 940 | 940 | 907 | 902 | 916 | 897 | 916 | 907 | 901 |
| 2 | 790 | 818 | 908 | 997 | 974 | 941 | 927 | 922 | 917 | 908 | 870 | 950 | 903 |
| 3 | 755 | 712 | 764 | 769 | 858 | 863 | 872 | 863 | 858 | 910 | 863 | 858 | 895 |
| 4 | 686 | 785 | 879 | 935 | 945 | 949 | 912 | 935 | 926 | 935 | 926 | 912 | 857 |
| 5 | 721 | 773 | 867 | 909 | 961 | 965 | 989 | 975 | 956 | 923 | 871 | 839 | 889 |
| 6 | 855 | 888 | 953 | 991 | 1024 | 930 | 756 | 770 | 737 | 775 | 765 | 855 | 892 |
| 7 | 744 | 767 | . . . | . . . | 1059 | 847 | 871 | 833 | 763 | 791 | 805 | 781 | 857 |
| 8 | 553 | 666 | 793 | 704 | 835 | 840 | 868 | 854 | 859 | 859 | 816 | 816 | 783 |
| 9 | 691 | 701 | 734 | 781 | 847 | 884 | 875 | 856 | 851 | 828 | 823 | 828 | 824 |
| 10 | 712 | 759 | 778 | 853 | 905 | 910 | 905 | 905 | 896 | 891 | 867 | 881 | 855 |
| 11 | 794 | 855 | 860 | 888 | 902 | 874 | 841 | 841 | 836 | 827 | 827 | 832 | 873 |
| 12 | 787 | 787 | 801 | 834 | 885 | 914 | 918 | 923 | 914 | 904 | 895 | 885 | 850 |
| 13 | 845 | 878 | 906 | . . . | 1019 | 930 | 887 | 897 | 897 | 892 | 883 | 911 | 900 |
| 14 | 875 | 880 | 889 | 885 | 701 | 711 | 537 | 706 | 777 | 810 | 819 | 833 | 818 |
| 15 | 717 | 688 | . . . | . . . | 843 | 815 | 796 | 693 | 670 | 749 | 740 | 740 | 800 |
| 16 | 809 | 818 | 743 | 832 | 796 | 800 | 838 | 805 | 816 | 806 | 806 | 787 | 788 |
| 17 | 584 | 654 | 682 | 757 | 740 | 627 | 942 | 575 | 731 | 661 | 778 | 825 | 768 |
| 18 | 688 | 646 | 561 | 613 | 661 | 708 | 736 | 694 | 864 | 780 | 747 | 794 | 693 |
| 19 | 713 | 671 | 657 | 643 | 700 | 564 | 714 | 620 | 654 | 753 | 635 | 715 | 740 |
| 20 | 596 | 648 | 727 | 784 | 813 | 841 | 804 | 743 | 805 | 781 | 772 | 776 | 740 |
| 21 | 682 | 677 | 733 | 809 | 824 | 767 | 852 | 697 | 750 | 750 | 787 | 787 | 771 |
| 22 | 697 | 707 | 678 | 744 | 872 | 759 | 778 | 698 | 478 | 539 | 610 | 770 | 746 |
| 23 | 650 | 594 | 683 | 918 | 1027 | 891 | 882 | 886 | 845 | 836 | 808 | 784 | 799 |
| 24 | 769 | 797 | 802 | 863 | 869 | 883 | 907 | 911 | 884 | 884 | 851 | 828 | 847 |
| 25 | 756 | 634 | 611 | 785 | 931 | 974 | 936 | 927 | 914 | 904 | 890 | 885 | 876 |
| 26 | 808 | 827 | 888 | 945 | 941 | 828 | 871 | 861 | 848 | 763 | 810 | 740 | 863 |
| 27 | 575 | 814 | 829 | 805 | 830 | 867 | 839 | 764 | 831 | 741 | 788 | 821 | 830 |
| 28 | 745 | 835 | 905 | 924 | 939 | 873 | 897 | 845 | 837 | 738 | 597 | 738 | 799 |
| Mean . | 744 | 756 | 790 | 837 | 880 | 846 | 852 | 818 | 823 | 816 | 806 | 824 | 827 |

TABLE VI—Continued—*Hori*

The figures given in the table are millionths of a dyne, which, added

.198000 +

| Day. | A. M. | | | | | | | | | | | |
|---|---|---|---|---|---|---|---|---|---|---|---|---|
| | 1. | 2. | 3. | 4. | 5. | 6. | 7. | 8. | 9. | 10. | 11. | 12. |
| 1 | 405 | 409 | 687 | 767 | 739 | 768 | 782 | 617 | 618 | 599 | 693 | 670 |
| 2 | 712 | 735 | 749 | 735 | 807 | 736 | 764 | 745 | 709 | 662 | 620 | 605 |
| 3 | 802 | 826 | 816 | 807 | 803 | 831 | 794 | 714 | 687 | 649 | 602 | 611 |
| 4 | 846 | 813 | 775 | 808 | 837 | 851 | 837 | 861 | 815 | 862 | 867 | 805 |
| 5 | 874 | 846 | 855 | 851 | 847 | 889 | 885 | 814 | 773 | 806 | 857 | 890 |
| 6 | 894 | 1017 | 819 | 866 | 820 | 703 | 905 | 698 | 628 | 206 | 586 | 652 |
| 7 | 853 | 708 | 567 | 708 | 680 | 685 | 798 | 821 | 705 | 710 | 705 | 611 |
| 8 | 802 | 943 | 797 | 821 | 817 | 841 | 826 | 794 | 719 | 644 | 649 | 649 |
| 9 | 789 | 813 | 836 | 789 | 837 | 828 | 832 | 771 | 702 | 683 | 674 | 683 |
| 10 | 847 | 856 | 861 | 861 | 885 | 857 | 843 | 815 | 858 | 722 | 637 | 651 |
| 11 | 842 | 828 | 847 | 861 | 862 | 881 | 876 | 843 | 774 | 774 | 779 | 713 |
| 12 | 1041 | 980 | 928 | 975 | 986 | 939 | 925 | 896 | 855 | 860 | 794 | 752 |
| 13 | 812 | 859 | 765 | 878 | 808 | 874 | 931 | 982 | 889 | 885 | 856 | 833 |
| 14 | 785 | 781 | 781 | 795 | 833 | 805 | 796 | 749 | 717 | 707 | 674 | 684 |
| 15 | 795 | 814 | 833 | 855 | 852 | 866 | 843 | 740 | 708 | 661 | 914 | 679 |
| 16 | 824 | 839 | 881 | 848 | 848 | 871 | 867 | 792 | 759 | 698 | 688 | 669 |
| 17 | 868 | 849 | 863 | 896 | 877 | 849 | 882 | 830 | 727 | 623 | 713 | 825 |
| 18 | 596 | 615 | 620 | 714 | 718 | 690 | 596 | 526 | 526 | 521 | 577 | . . . |
| 19 | 776 | 804 | 799 | 813 | 827 | 837 | 813 | 780 | 729 | 710 | 691 | 691 |
| 20 | 786 | 819 | 805 | 819 | 814 | 838 | 791 | 781 | 735 | 617 | 565 | 655 |
| 21 | 768 | 862 | 815 | 825 | 834 | 834 | 834 | 754 | 726 | 693 | 651 | 552 |
| 22 | 953 | 901 | 713 | 661 | 798 | 774 | 802 | 760 | 708 | 567 | 572 | 581 |
| 23 | 761 | 780 | 846 | 789 | 799 | 761 | 756 | 799 | 728 | 662 | 695 | 761 |
| 24 | 800 | 879 | 832 | 832 | 851 | 738 | 785 | 814 | 795 | 781 | 677 | 649 |
| 25 | 768 | 744 | 768 | 805 | 791 | 791 | 772 | 711 | 716 | 763 | 739 | 697 |
| 26 | 863 | 816 | 830 | 830 | 830 | 820 | 830 | 797 | 736 | 712 | 712 | 750 |
| 27 | 713 | 765 | 737 | 732 | 788 | 774 | 784 | 680 | 619 | 718 | 788 | 793 |
| 28 | 451 | 587 | 587 | 559 | 677 | 695 | 438 | 818 | 625 | 545 | 503 | 498 |
| 29 | 701 | 790 | 701 | 701 | 711 | 743 | 772 | 645 | 518 | 452 | 438 | 368 |
| 30 | 730 | 796 | 796 | 806 | 801 | 782 | 796 | 735 | 650 | 500 | 486 | 509 |
| 31 | 830 | 807 | 830 | 802 | 778 | 816 | 807 | 727 | 609 | 557 | 595 | 628 |
| Mean . | 783 | 803 | 785 | 800 | 811 | 805 | 807 | 768 | 712 | 663 | 679 | 670 |

*zontal force March,* 1889.

to .198000 dyne, give the absolute horizontal force in C. G. S. units.

.198000 +

| Day. | P. M. | | | | | | | | | | | | Mean. |
|---|---|---|---|---|---|---|---|---|---|---|---|---|---|
| | 1. | 2. | 3. | 4. | 5. | 6. | 7. | 8. | 9. | 10. | 11. | 12. | |
| 1 | 610 | 563 | 699 | 699 | 794 | 738 | 747 | 757 | 706 | 678 | 729 | 729 | 675 |
| 2 | 658 | 691 | 550 | 696 | 786 | 824 | 819 | 786 | 782 | 782 | 792 | 787 | 731 |
| 3 | 655 | 772 | 871 | 833 | 858 | 839 | 839 | 839 | 849 | 779 | 798 | 845 | 780 |
| 4 | 830 | 821 | 882 | .900 | 939 | 930 | 925 | 831 | 846 | 832 | 818 | 921 | 852 |
| 5 | 858 | 896 | 915 | 967 | 935 | 1015 | 968 | 953 | 743 | 419 | 541 | 658 | 836 |
| 6 | 653 | 488 | 653 | 761 | 814 | 814 | 786 | 804 | 801 | 772 | 707 | 594 | 727 |
| 7 | 668 | 654 | 664 | 800 | 707 | 655 | 829 | 853 | 802 | 778 | 882 | 755 | 733 |
| 8 | 739 | 913 | 927 | 847 | 797 | 768 | 768 | 815 | 802 | 816 | 807 | 793 | 796 |
| 9 | 689 | 717 | 773 | 825 | 845 | 859 | 864 | 845 | 865 | 879 | 851 | 851 | 796 |
| 10 | 667 | 690 | 775 | 840 | 856 | 865 | 870 | 856 | 852 | 861 | 857 | 838 | 813 |
| 11 | 845 | 892 | 939 | 902 | 908 | 926 | 922 | 922 | 913 | 890 | 866 | 866 | 861 |
| 12 | 790 | 880 | 983 | 1007 | 904 | 636 | 730 | 918 | 919 | 872 | 863 | 849 | 887 |
| 13 | 857 | 872 | 716 | 669 | 638 | 741 | 652 | 595 | 507 | 671 | 671 | 624 | 774 |
| 14 | 755 | 779 | 784 | 812 | 836 | 813 | 747 | 799 | 828 | 833 | 800 | 889 | 783 |
| 15 | 690 | 798 | 727 | 831 | 851 | 860 | 888 | 884 | 842 | 828 | 838 | 828 | 809 |
| 16 | 787 | 853 | 890 | 876 | 857 | 829 | 806 | 857 | 881 | 890 | 900 | 895 | 829 |
| 17 | 835 | 821 | 1206 | 835 | 445 | −185 | 149 | −039 | 356 | 426 | 454 | 595 | 658 |
| 18 | 662 | 714 | 784 | 841 | 855 | 831 | 850 | 855 | 826 | 737 | 714 | 789 | 702 |
| 19 | 701 | 771 | 799 | 827 | 823 | 715 | 818 | 809 | 809 | 804 | 846 | 804 | 783 |
| 20 | 702 | 725 | 739 | 739 | 706 | 810 | 749 | 777 | 758 | 739 | 711 | 753 | 747 |
| 21 | 585 | 731 | 717 | 712 | 811 | 717 | 820 | 839 | 829 | 858 | 938 | 905 | 775 |
| 22 | 605 | 732 | 793 | 760 | 647 | 624 | 628 | 671 | 713 | 802 | 779 | 793 | 722 |
| 23 | 799 | 803 | 817 | 831 | 822 | 836 | 860 | 869 | 836 | 822 | 822 | 784 | 793 |
| 24 | 743 | 800 | 847 | 879 | 922 | 818 | 832 | 842 | 785 | 724 | 738 | 842 | 800 |
| 25 | 749 | 782 | 848 | 843 | 815 | 801 | 791 | 805 | 815 | 796 | 815 | 824 | 781 |
| 26 | 900 | 947 | 924 | 900 | 858 | 830 | 708 | 731 | 722 | 698 | .... | .... | 806 |
| 27 | 962 | 793 | 864 | 915 | 925 | 859 | 864 | 835 | 751 | 605 | 976 | 643 | 787 |
| 28 | 822 | 493 | 757 | 719 | 738 | 808 | 686 | 677 | 719 | 719 | 639 | 625 | 644 |
| 29 | 725 | 762 | 767 | 711 | 762 | 711 | 795 | 814 | 823 | 828 | 837 | 847 | 705 |
| 30 | 524 | 538 | 608 | 688 | 768 | 754 | 782 | 796 | 810 | 796 | 801 | 806 | 711 |
| 31 | 633 | 741 | 830 | 835 | 811 | 816 | 821 | 830 | 830 | 816 | 783 | 797 | 764 |
| Mean . | 726 | 756 | 808 | 816 | 808 | 766 | 784 | 788 | 785 | 766 | 786 | 784 | 769 |

TABLE VI—Continued—*Hori*

The figures given in the table are millionths of a dyne, which, added

.198000 +

| Day. | A. M. | | | | | | | | | | | |
|---|---|---|---|---|---|---|---|---|---|---|---|---|
| | 1. | 2. | 3. | 4. | 5. | 6. | 7. | 8. | 9. | 10. | 11. | 12. |
| 1 | 818 | 832 | 879 | 846 | 874 | 856 | 856 | 856 | 809 | 682 | 611 | 630 |
| 2 | 631 | 777 | 786 | 833 | 871 | 922 | 871 | 711 | 607 | 570 | 584 | 650 |
| 3 | 920 | 1032 | 750 | 981 | 882 | 901 | 830 | 840 | 821 | 689 | 567 | 548 |
| 4 | 751 | 770 | 784 | 789 | 808 | 798 | 780 | 700. | 610 | 535 | 493 | 423 |
| 5 | 795 | 785 | 799 | 799 | 799 | 818 | 795 | 762 | 682 | 611 | 560 | 588 |
| 6 | 810 | 805 | 805 | 810 | 824 | 824 | 810 | 749 | 721 | 739 | 810 | 833 |
| 7 | 919 | 905 | 910 | 858 | 905 | 872 | 783 | 787 | 754 | 665 | 693 | 679 |
| 8 | 643 | 831 | 666 | 572 | 666 | 619 | 591 | 525 | 450 | 323 | 488 | 671 |
| 9 | 667 | 714 | 695 | 719 | 719 | 695 | 658 | 634 | 573 | 437 | 399 | 470 |
| 10 | 602 | 626 | 649 | 555 | 715 | 631 | 607 | 518 | 574 | 579 | 612 | 682 |
| 11 | 683 | 679 | 650 | 702 | 711 | 740 | 758 | 740 | 693 | 707 | 791 | 848 |
| 12 | 741 | 745 | 750 | 750 | 759 | 783 | 759 | 703 | 623 | 614 | 661 | 722 |
| 13 | 713 | 742 | 751 | 751 | 770 | 760 | 695 | 666 | 591 | 911 | 671 | 652 |
| 14 | 790 | 771 | 771 | 790 | 790 | 808 | 752 | 691 | 639 | 700 | 663 | 639 |
| 15 | 777 | 772 | 730 | 748 | 767 | 791 | 762 | 711 | 626 | 607 | 589 | 626 |
| 16 | 792 | 796 | 820 | 820 | 843 | 876 | 810 | 698 | 580 | 528 | 575 | 674 |
| 17 | 770 | 808 | 817 | 817 | 845 | 855 | 822 | 690 | 629 | 601 | 653 | 794 |
| 18 | 837 | 818 | 832 | 832 | 804 | 842 | 781 | 658 | 503 | 503 | 611 | 677 |
| 19 | 814 | 782 | 763 | 791 | 791 | 791 | 749 | 631 | 509 | 471 | 457 | 500 |
| 20 | 867 | 806 | 787 | 783 | 787 | 787 | 764 | 679 | 585 | 562 | 642 | 764 |
| 21 | 855 | 775 | 794 | 780 | 813 | 813 | 817 | 770 | 681 | 597 | 615 | 775 |
| 22 | 818 | 804 | 837 | 842 | 823 | 814 | 776 | 743 | 687 | 692 | 692 | 687 |
| 23 | 810 | 801 | 834 | 819 | 857 | 805 | 782 | 655 | 613 | 655 | 674 | 782 |
| 24 | 741 | 759 | 773 | 773 | 778 | 769 | 731 | 726 | 679 | 647 | 656 | 726 |
| 25 | 821 | 850 | 878 | 845 | 746 | 850 | 732 | 756 | 666 | 765 | 793 | 864 |
| 26 | 800 | 706 | 673 | 635 | 706 | 701 | 650 | 537 | 443 | 339 | 424 | 574 |
| 27 | 749 | 773 | 777 | 707 | 726 | 759 | 763 | 651 | 481 | 364 | 416 | 369 |
| 28 | 769 | 769 | 811 | 778 | 830 | 854 | 760 | 689 | 572 | 553 | 605 | 652 |
| 29 | 808 | 789 | 779 | 784 | 798 | 826 | 808 | 695 | 554 | 483 | 587 | 728 |
| 30 | 823 | 842 | 813 | 837 | 799 | 762 | 743 | 691 | 696 | 696 | 748 | 860 |
| Mean . | 778 | 785 | 779 | 778 | 794 | 797 | 757 | 709 | 622 | 594 | 611 | 670 |

*zontal force, April,* 1889.

to .198000 dyne, give the absolute horizontal force in C. G. S. units.

.198000 +

| Day. | P. M. | | | | | | | | | | | | Mean. |
|---|---|---|---|---|---|---|---|---|---|---|---|---|---|
| | 1. | 2. | 3. | 4. | 5. | 6. | 7. | 8. | 9. | 10. | 11. | 12. | |
| 1 | 682 | 752 | 827 | 724 | 719 | 785 | 766 | 757 | 785 | 809 | 700 | 809 | 778 |
| 2 | 777 | 748 | 814 | 842 | 753 | 828 | 791 | 786 | 847 | 758 | 814 | 847 | 767 |
| 3 | 567 | 619 | 750 | 741 | 849 | 722 | 732 | 755 | 755 | 718 | 746 | 736 | 769 |
| 4 | 592 | 751. | 780 | 808 | 808 | 770 | 789 | 798 | 803 | 803 | 803 | 798 | 735 |
| 5 | 616 | 602 | 644 | 767 | 795 | 814 | 818 | 828 | 795 | 790 | 790 | 799 | 744 |
| 6 | 815 | 810 | 768 | 768 | 805 | 758 | 800 | 810 | 838 | 885 | 909 | 937 | 810 |
| 7 | 750 | 707 | 801 | 740 | 830 | 750 | 801 | 745 | 449 | —021 | 482 | 576 | 724 |
| 8 | 845 | 798 | 516 | 488 | 657 | 633 | 633 | 676 | 525 | 614 | 666 | 643 | 614 |
| 9 | 545 | 686 | 850 | 756 | 630 | 672 | 390 | 592 | 615 | 446 | 536 | 507 | 609 |
| 10 | 795 | 800 | 809 | 710 | 682 | 710 | 725 | 715 | 710 | 729 | 663 | 645 | 668 |
| 11 | 862 | 848 | 801 | 730 | 810 | 791 | 693 | 735 | 730 | 735 | 744 | 749 | 747 |
| 12 | 835 | 839 | 830 | 896 | 811 | 665 | 689 | 689 | 689 | 680 | 670 | 896 | 742 |
| 13 | 770 | 817 | 737 | 775 | 793 | 779 | 789 | 807 | 760 | 760 | 746 | 770 | 749 |
| 14 | 761 | 846 | 785 | 808 | 823 | 738 | 682 | 696 | 757 | 785 | 785 | 785 | 752 |
| 15 | 673 | 809 | 856 | 819 | 795 | 781 | 739 | 758 | 777 | 786 | 772 | 762 | 743 |
| 16 | 703 | 844 | 858 | 877 | 863 | 774 | 797 | 774 | 783 | 732 | 783 | 774 | 766 |
| 17 | 798 | 822 | 803 | 780 | 780 | 775 | 775 | 803 | 827 | 756 | 737 | 869 | 776 |
| 18 | 715 | 748 | 804 | 818 | 884 | 799 | 790 | 790 | 743 | 776 | 771 | 799 | 756 |
| 19 | 514 | 565 | 655 | 749 | 810 | 838 | 838 | 847 | 824 | 782 | 669 | 716 | 702 |
| 20 | 840 | 849 | 943 | 925 | 915 | 901 | 760 | 760 | 718 | 732 | 713 | 774 | 777 |
| 21 | 742 | 827 | 846 | 813 | 752 | 808 | 799 | 836 | 761 | 879 | 832 | 728 | 780 |
| 22 | 701 | 753 | 739 | 790 | 804 | 814 | 776 | 748 | 837 | 724 | 771 | 781 | 769 |
| 23 | 909 | 862 | 885 | 758 | 707 | 617 | 542 | 669 | 721 | 758 | 542 | 693 | 740 |
| 24 | 708 | 764 | 788 | 820 | 820 | 820 | 802 | 835 | 835 | 839 | 830 | 825 | 769 |
| 25 | 888 | 954 | 696 | 733 | 822 | 564 | ·587 | 752 | 775 | 700 | 658 | 916 | 775 |
| 26 | 701 | 772 | 791 | 715 | 762 | 715 | 729 | 748 | 758 | 739 | 739 | 734 | 670 |
| 27 | 519 | 745 | 787 | 829 | 853 | . . . | 792 | 712 | 740 | 594 | 449 | 632 | 660 |
| 28 | 746 | 797 | 821 | 774 | 807 | 699 | 713 | 717 | 778 | 764 | 976 | 825 | 752 |
| 29 | 765 | 747 | 822 | 756 | 808 | 826 | 784 | 784 | 709 | 751 | 803 | 836 | 751 |
| 30 | 889 | 912 | 964 | 818 | 771 | 743 | 743 | 771 | 780 | 705 | 780 | 752 | 789 |
| Mean . | 734 | 780 | 792 | 778 | 791 | 755 | 735 | 756 | 747 | 717 | 729 | 764 | 740 |

TABLE VI—Continued—*Hor*

The figures given in the table are millionths of a dyne, which, added

.198000 +

| Day. | A. M. | | | | | | | | | | | |
|---|---|---|---|---|---|---|---|---|---|---|---|---|
| | 1. | 2. | 3. | 4. | 5. | 6. | 7. | 8. | 9. | 10. | 11. | 12. |
| 1 | 777 | 791 | 791 | 753 | 781 | 796 | 763 | 683 | 546 | 504 | 584 | 711 |
| 2 | 764 | 768 | 787 | 782 | 778 | 797 | 768 | 674 | 547 | 468 | 463 | 580 |
| 3 | 798 | 798 | 807 | 812 | 816 | 840 | 802 | 704 | 586 | 548 | 501 | 595 |
| 4 | 620 | 672 | 620 | 667 | 761 | 761 | 761 | 766 | 648 | 474 | 498 | 596 |
| 5 | 908 | 668 | 626 | 663 | 757 | 743 | 673 | 729 | 626 | 438 | 461 | 602 |
| 6 | 674 | 627 | 655 | 660 | 678 | 678 | 589 | 528 | 434 | 448 | 561 | 645 |
| 7 | 482 | 458 | 552 | 599 | 599 | 642 | 595 | 482 | 327 | 379 | 567 | 703 |
| 8 | 677 | 630 | 648 | 667 | 677 | 700 | 658 | 507 | 404 | 385 | 559 | 667 |
| 9 | 711 | 753 | 776 | 729 | 767 | 711 | 715 | 621 | 555 | 551 | 668 | 781 |
| 10 | 721 | 730 | 740 | 806 | 735 | 730 | 716 | 702 | 519 | 500 | 608 | 716 |
| 11 | 680 | 675 | 656 | 642 | 647 | 713 | 698 | 595 | 478 | 492 | 562 | . . . |
| 12 | 718 | 699 | 718 | 751 | 775 | 765 | 746 | 685 | 634 | 643 | 826 | 977 |
| 13 | 710 | 715 | 762 | 744 | 720 | 734 | 684 | 552 | 491 | 487 | 666 | 807 |
| 14 | 695 | 733 | 667 | 687 | 696 | 734 | 707 | 575 | 411 | 299 | 426 | 651 |
| 15 | 690 | 681 | 690 | 682 | 710 | 719 | 702 | 622 | 401 | 341 | 472 | 637 |
| 16 | 700 | 733 | 710 | 720 | 739 | 786 | 792 | 651 | 355 | 219 | 360 | 506 |
| 17 | 728 | 742 | 747 | 743 | 771 | 790 | 777 | 641 | 453 | 426 | 548 | 642 |
| 18 | 737 | 751 | 761 | 776 | 794 | 832 | 791 | 603 | 391 | 336 | 449 | 547 |
| 19 | 658 | 724 | 672 | 645 | 645 | 687 | 702 | 655 | 547 | 482 | 473 | 515 |
| 20 | 658 | 658 | 649 | 650 | 683 | 630 | 693 | 627 | 543 | 516 | 534 | 633 |
| 21 | 723 | 699 | 723 | 700 | 700 | 672 | 654 | 621 | 635 | 740 | 810 | 782 |
| 22 | . . . | . . . | . . . | | | . . . | | . . . | . . . | 392 | 439 | 486 |
| 23 | 647 | 614 | 623 | 643 | 681 | 709 | 672 | 559 | 409 | 325 | 382 | 518 |
| 24 | 646 | 646 | 609 | 629 | 713 | 737 | 648 | 442 | 404 | 419 | 541 | 626 |
| 25 | 675 | 666 | 675 | 681 | ·695 | 718 | 710 | 607 | 517 | 471 | 551 | 659 |
| 26 | 839 | 1117 | 905 | 836 | 779 | 695 | 461 | 710 | 559 | 532 | 419 | 438 |
| 27 | 627 | 679 | 623 | 600 | 619 | 652 | 690 | 545 | 390 | 344 | 522 | 635 |
| 28 | 650 | 679 | 693 | 651 | 619 | 651 | 610 | 591 | 455 | 475 | 489 | 536 |
| 29 | 646 | 646 | 646 | 624 | 694 | 699 | 742 | 714 | 639 | 504 | 541 | 649 |
| 30 | 655 | 613 | 651 | 680 | 708 | 746 | 765 | 709 | 657 | 555 | 602 | 668 |
| 31 | 627 | 664 | 716 | 689 | 665 | 670 | 535 | 544 | 446 | 432 | 522 | 649 |
| Mean . | 695 | 701 | 697 | 697 | 713 | 725 | 694 | 621 | 500 | 456 | 536 | 639 |

*zontal force, May,* 1889.

to .198000 dyne, give the absolute horizontal force in C. G. S. units.

.198000 +

| Day. | P. M. | | | | | | | | | | | | Mean. |
| | 1. | 2. | 3. | 4. | 5. | 6. | 7. | 8. | 9. | 10. | 11. | 12. | |
|---|---|---|---|---|---|---|---|---|---|---|---|---|---|
| 1 | 763 | 894 | 880 | 725 | 725 | . . . | 786 | 796 | 800 | 847 | 781 | 777 | 750 |
| 2 | 693 | 731 | 787 | 792 | 778 | 764 | 811 | 806 | 820 | 815 | 815 | 801 | 733 |
| 3 | 713 | 826 | 915 | 957 | 962 | 948 | 906 | 812 | 877 | 779 | 741 | 769 | 784 |
| 4 | 742 | 709 | 733 | 653 | 770 | 808 | 789 | 789 | 784 | 780 | 761 | 723 | 704 |
| 5 | 706 | 767 | 673 | 626 | 550 | 673 | 654 | 659 | 659 | 654 | 630 | 668 | 659 |
| 6 | 791 | 895 | 895 | 824 | 805 | 744 | 749 | 721 | 692 | 735 | 923 | 622 | 691 |
| 7 | 792 | 896 | 830 | 693 | 722 | 708 | 712 | 731 | 698 | 637 | 614 | 731 | 635 |
| 8 | 845 | 854 | 831 | 765 | 694 | 666 | 699 | 709 | 718 | 727 | 718 | 709 | 671 |
| 9 | 856 | 880 | 880 | 772 | 809 | 753 | 720 | 725 | 649 | 645 | 664 | 659 | 723 |
| 10 | 655 | 594 | 542 | 618 | 683 | 693 | 712 | 693 | 707 | 697 | 669 | 693 | 674 |
| 11 | 821 | 825 | 807 | 774 | 741 | 684 | 722 | 760 | 764 | 731 | 727 | 722 | 692 |
| 12 | 1118 | 1033 | 906 | 883 | 906 | 718 | 784 | 657 | 662 | 779 | 873 | 850 | 796 |
| 13 | 907 | 874 | 850 | 682 | 705 | 687 | 665 | 684 | 684 | 676 | 685 | 685 | 702 |
| 14 | 817 | 836 | 831 | 658 | 667 | 644 | 674 | 674 | 669 | 680 | 680 | 689 | 658 |
| 15 | 785 | 897 | 869 | 790 | 739 | 715 | 694 | 675 | 717 | 713 | 699 | 718 | 682 |
| 16 | 685 | 761 | 826 | 846 | 809 | 776 | 717 | 679 | 717 | 755 | 751 | 718 | 680 |
| 17 | 737 | 770 | 788 | 766 | 799 | 757 | 768 | 754 | 726 | 722 | 722 | 722 | 710 |
| 18 | 628 | 783 | 783 | 775 | 775 | 742 | 730 | 697 | 711 | 726 | 750 | 726 | 691 |
| 19 | 597 | 635 | 677 | 711 | 659 | 687 | 661 | 656 | 609 | 643 | 667 | 634 | 635 |
| 20 | 742 | 761 | 714 | 743 | 743 | 715 | 712 | 708 | 698 | 704 | 685 | 666 | 669 |
| 21 | . . . | . . . | . . . | . . . | . . . | . . . | . . . | . . . | . . . | . . . | . . . | . . . | 705 |
| 22 | 529 | 548 | 623 | . . . | . . . | . . . | 523 | 598 | 555 | 542 | 571 | 594 | 533 |
| 23 | 768 | 834 | 834 | 765 | 689 | 703 | 710 | 673 | 644 | 678 | 678 | 664 | 643 |
| 24 | 655 | 66. | 702 | 797 | 731 | 703 | 663 | 780 | 663 | 687 | 673 | 654 | 643 |
| 25 | 797 | 891 | 839 | 802 | 779 | 755 | 748 | 743 | 753 | 768 | 768 | 749 | 709 |
| 26 | 533 | 754 | 716 | 355 | 567 | 572 | 508 | 428 | 287 | 448 | 476 | 537 | 603 |
| 27 | 678 | 660 | 702 | 693 | 637 | 628 | 658 | 601 | 620 | 621 | 645 | 635 | 613 |
| 28 | 702 | 825 | 843 | 802 | 732 | 656 | 663 | 677 | 658 | 650 | 664 | 645 | 651 |
| 29 | 758 | 805 | 899 | 835 | 759 | 722 | 696 | 686 | 705 | 706 | 692 | 678 | 695 |
| 30 | 772 | 913 | 871 | 848 | 689 | 416 | 470 | 681 | 780 | 428 | 612 | 621 | 671 |
| 31 | 664 | 711 | 617 | 599 | 618 | 580 | 620 | 638 | 671 | 635 | 668 | 654 | 618 |
| Mean . | 742 | 794 | 789 | 743 | 734 | 701 | 697 | 696 | 690 | 687 | 700 | 690 | 678 |

TABLE VI—Continued—*Hori*

The figures given in the table are millionths of a dyne, which, added

.198000 +

| Day. | A. M. | | | | | | | | | | | |
|------|------|------|------|------|------|------|------|------|------|------|------|------|
| | 1. | 2. | 3. | 4. | 5. | 6. | 7. | 8. | 9. | 10. | 11. | 12. |
| 1 | 683 | 763 | 622 | 646 | 627 | 674 | 634 | 564 | 559 | 541 | 466 | . . . |
| 2 | 674 | 683 | 674 | 642 | 661 | 689 | 610 | 577 | 450 | 301 | 339 | 428 |
| 3 | 739 | 692 | 664 | 665 | 656 | 665 | 699 | 596 | 403 | 362 | 498 | 653 |
| 4 | 669 | 669 | 687 | 641 | 674 | 702 | 609 | 497 | 450 | 460 | 592 | 695 |
| 5 | 673 | 729 | 682 | 693 | 679 | 711 | 662 | 544 | 441 | 418 | 526 | 653 |
| 6 | 603 | 626 | 678 | 693 | 712 | 707 | 713 | 699 | 643 | 770 | 676 | 681 |
| 7 | 668 | 650 | 650 | 636 | 651 | 693 | 689 | 633 | 586 | 587 | 728 | 826 |
| 8 | 720 | 729 | 748 | 758 | 749 | 754 | 708 | 642 | 562 | 511 | 511 | 577 |
| 9 | 827 | 804 | 766 | 570 | 687 | 711 | 552 | 604 | 496 | 356 | 407 | 478 |
| 10 | 606 | 639 | 658 | 645 | 682 | 678 | 679 | 688 | 627 | 576 | 604 | 694 |
| 11 | 671 | 634 | 723 | 682 | 677 | 677 | 692 | 636 | 584 | 524 | 505 | 477 |
| 12 | 633 | 648 | 671 | 663 | 672 | 677 | 706 | 673 | 664 | 665 | 632 | 613 |
| 13 | 699 | 736 | 751 | 737 | 789 | 799 | 707 | 725 | 650 | 609 | 618 | 755 |
| 14 | 497 | 610 | 408 | 555 | 630 | 423 | 330 | 161 | 137 | 308 | 289 | 383 |
| 15 | 389 | 431 | 464 | 474 | 521 | 498 | 471 | 395 | 297 | 227 | 321 | 486 |
| 16 | 567 | 544 | 567 | 592 | 629 | 639 | 546 | . . . | . . . | . . . | . . . | . . . |
| 17 | 543 | 507 | 577 | 569 | 649 | 664 | 651 | 580 | 482 | 482 | 559 | 554 |
| 18 | 609 | 628 | 675 | 672 | 672 | 738 | 712 | 641 | 580 | 534 | 506 | 595 |
| 19 | 642 | 665 | 660 | 686 | 733 | 733 | 726 | 585 | 566 | 524 | 548 | 618 |
| 20 | 661 | 666 | 685 | 673 | 692 | 692 | 656 | 590 | 576 | 502 | 601 | 690 |
| 21 | 628 | 755 | 774 | 649 | 607 | 701 | 661 | 623 | 585 | 572 | 756 | 633 |
| 22 | 675 | 586 | 619 | 640 | 677 | 626 | 599 | 609 | 477 | 351 | 328 | 332 |
| 23 | 661 | 628 | 628 | 546 | 635 | 663 | 623 | 604 | 510 | 506 | 478 | 506 |
| 24 | 634 | 643 | 667 | 664 | 664 | 692 | 614 | 464 | 412 | 366 | 498 | 601 |
| 25 | 699 | 629 | 610 | 654 | 659 | 673 | 633 | 558 | 445 | 319 | 357 | 488 |
| 26 | 671 | 671 | 643 | 673 | 678 | 697 | 722 | 605 | 464 | 310 | 366 | 446 |
| 27 | 680 | 695 | 695 | 687 | 697 | 715 | 684 | 572 | 435 | 432 | 587 | 765 |
| 28 | 757 | 728 | 827 | 721 | 768 | 857 | 770 | 671 | 507 | 428 | . . . | . . . |
| 29 | 615 | 592 | 592 | 617 | 660 | 655 | 619 | 521 | 361 | 296 | 301 | 423 |
| 30 | 634 | 601 | 611 | 627 | 631 | 674 | 662 | 554 | 441 | 399 | 437 | 498 |
| Mean . | 648 | 653 | 656 | 646 | 670 | 682 | 644 | 580 | 497 | 456 | 500 | 576 |

*zontal force, June,* 1889.

to .198000 dyne, give the absolute horizontal force in C. G. S. units.

.198000 +

| Day. | P. M. | | | | | | | | | | | | Mean. |
|---|---|---|---|---|---|---|---|---|---|---|---|---|---|
| | 1. | 2. | 3. | 4. | 5. | 6. | 7. | 8. | 9. | 10. | 11. | 12. | |
| 1 | 448 | 519 | 646 | 665 | 665 | 614 | 634 | 705 | 639 | 748 | 650 | 659 | 625 |
| 2 | 593 | 711 | 720 | 726 | 670 | 665 | 681 | 681 | 705 | 710 | 691 | 626 | 621 |
| 3 | 790 | 776 | 781 | 820 | 749 | 561 | 652 | 685 | 704 | 733 | 743 | 686 | 666 |
| 4 | 710 | 762 | 799 | 805 | 768 | 674 | 661 | 647 | 666 | 677 | 677 | 672 | 661 |
| 5 | 795 | 913 | 918 | 796 | 716 | 646 | 648 | 695 | 704 | 677 | 696 | 649 | 678 |
| 6 | 781 | 748 | 757 | 735 | 711 | 716 | 742 | 666 | 671 | 714 | 649 | 681 | 699 |
| 7 | 907 | 936 | 851 | 796 | 687 | 683 | 713 | 713 | 727 | 737 | 737 | 719 | 713 |
| 8 | 606 | 695 | 771 | 946 | 823 | 884 | 896 | 910 | 886 | 944 | 953 | 991 | 761 |
| 9 | 531 | 573 | 620 | 677 | 682 | 696 | 632 | 628 | 576 | 614 | 600 | 605 | 612 |
| 10 | 695 | 699 | 629 | 724 | 766 | 780 | 731 | 688 | 731 | 685 | 656 | 652 | 676 |
| 11 | 577 | 694 | 713 | 785 | 714 | 653 | 641 | 631 | 655 | 661 | 661 | 665 | 647 |
| 12 | 661 | 722 | 713 | 685 | 662 | 662 | 631 | 650 | 687 | 698 | 698 | 735 | 672 |
| 13 | . . . | . . . | . . . | . . . | . . . | 747 | 749 | 716 | 754 | 689 | 882 | 473 | 715 |
| 14 | 483 | 548 | 525 | 502 | 333 | 314 | 255 | 293 | 377 | 393 | 383 | 444 | 399 |
| 15 | 675 | 736 | 708 | 657 | 671 | 648 | 593 | 532 | 471 | 580 | 519 | 467 | 506 |
| 16 | 524 | 642 | 787 | 624 | 591 | 591 | 583 | 569 | 739 | 472 | 552 | 552 | 595 |
| 17 | 639 | 766 | 785 | 825 | 651 | 618 | 653 | 653 | 653 | 659 | 607 | 583 | 621 |
| 18 | 691 | 804 | 874 | 900 | 707 | 684 | 662 | 681 | 681 | 654 | 635 | 621 | 673 |
| 19 | 653 | 672 | 639 | 707 | 608 | 665 | 733 | 662 | 643 | 603 | 626 | 636 | 647 |
| 20 | 739 | 781 | 706 | 821 | 919 | 628 | 705 | 743 | 691 | 739 | 734 | 687 | 691 |
| 21 | . . . | 729 | 706 | 694 | 567 | 642 | 682 | 583 | 653 | 438 | 631 | 584 | 644 |
| 22 | 598 | 706 | 663 | 661 | 665 | 651 | 630 | 592 | 625 | 645 | 612 | 631 | 592 |
| 23 | 574 | 579 | 630 | 632 | 637 | 637 | 620 | 630 | 696 | 641 | 622 | 646 | 605 |
| 24 | 617 | 702 | 688 | 756 | 746 | 713 | 682 | 739 | 701 | 777 | 636 | 627 | 638 |
| 25 | 575 | 702 | 725 | 652 | 605 | 600 | 621 | 640 | 668 | 646 | 679 | 664 | 604 |
| 26 | 617 | 655 | 645 | 675 | 661 | 652 | 682 | 677 | 692 | 704 | 699 | 680 | 620 |
| 27 | 861 | 904 | 890 | 821 | 760 | 713 | 753 | 809 | 837 | 858 | 863 | 820 | 731 |
| 28 | 552 | 670 | 900 | 516 | 493 | 592 | 401 | 396 | 448 | 463 | 552 | 642 | 621 |
| 29 | 585 | 726 | 655 | 714 | 634 | 653 | 641 | 584 | 636 | 689 | 660 | 632 | 582 |
| 30 | 613 | 651 | 660 | 704 | 629 | 653 | 678 | 697 | 697 | 671 | 708 | 798 | 622 |
| Mean . | 648 | 714 | 728 | 725 | 672 | 655 | 653 | 650 | 667 | 664 | 667 | 651 | 638 |

9230——4

TABLE VI—Continued—*Hori*

The figures given in the table are millionths of a dyne, which, added

.198000 +

| Day. | A. M. | | | | | | | | | | | |
|---|---|---|---|---|---|---|---|---|---|---|---|---|
| | 1. | 2. | 3. | 4. | 5. | 6. | 7. | 8. | 9. | 10. | 11. | 12. |
| 1 | 801 | 970 | 778 | 661 | 544 | 732 | 663 | 587 | 480 | 400 | 341 | 341 |
| 2 | 697 | 796 | 764 | 731 | 690 | 629 | 714 | 667 | 508 | 320 | 407 | 529 |
| 3 | 786 | 589 | 637 | 628 | 624 | 647 | 587 | 493 | 433 | 424 | 435 | 416 |
| 4 | 617 | 622 | 632 | 656 | 680 | 727 | 723 | 827 | 762 | 518 | 562 | 567 |
| 5 | 712 | 773 | 595 | 633 | 690 | 747 | 710 | 602 | 509 | 561 | 577 | 629 |
| 6 | 627 | 533 | 760 | 694 | 620 | 582 | 513 | 729 | 753 | 692 | 657 | 755 |
| 7 | 463 | 477 | 887 | 591 | 587 | 606 | 738 | 771 | 847 | 767 | 746 | . . . |
| 8 | . . . | . . . | . . . | . . . | . . . | . . . | . . . | . . . | . . . | . . . | 511 | 534 |
| 9 | 656 | 656 | 667 | 634 | 644 | 611 | 664 | 636 | 571 | 444 | 465 | 573 |
| 10 | 651 | 656 | 681 | 676 | 672 | 686 | 631 | 513 | 392 | 359 | 460 | 624 |
| 11 | 679 | 679 | 685 | 666 | 667 | 705 | 664 | 527 | 453 | 486 | 473 | 581 |
| 12 | 646 | 642 | 633 | 605 | 630 | 620 | 546 | 396 | 289 | 227 | 281 | 441 |
| 13 | 643 | 638 | 630 | 639 | 645 | 663 | 646 | 547 | 407 | 332 | 306 | 451 |
| 14 | 628 | 619 | 625 | 639 | 616 | 616 | 622 | 537 | 416 | 209 | 249 | 385 |
| 15 | 539 | 544 | 587 | 592 | 640 | 663 | 641 | 598 | 505 | 411 | 413 | 460 |
| 16 | 614 | 643 | 658 | 674 | 609 | 643 | 620 | 513 | 382 | 304 | 370 | 405 |
| 17 | 1058 | 754 | 849 | 342 | 658 | 485 | 378 | 473 | 465 | 315 | 251 | 115 |
| 18 | 370 | 521 | 569 | 458 | 482 | 516 | 508 | 269 | 355 | 360 | 263 | 296 |
| 19 | 457 | 627 | 483 | 507 | 518 | 514 | 496 | 431 | 375 | 471 | 274 | 304 |
| 20 | 591 | 541 | 603 | 538 | 567 | 568 | 508 | 377 | 327 | 398 | 399 | 400 |
| 21 | 496 | 525 | 531 | 560 | 566 | 567 | 493 | 451 | 349 | 341 | 384 | 441 |
| 22 | 536 | 532 | 576 | 600 | 564 | 598 | 547 | 449 | 389 | 399 | 419 | 514 |
| 23 | 633 | 605 | 588 | 546 | 590 | 624 | 629 | 630 | 584 | 501 | 600 | 573 |
| 24 | 701 | 716 | 670 | 718 | 686 | 682 | 674 | 567 | 432 | 386 | 448 | 585 |
| 25 | 638 | 657 | 658 | 655 | 665 | 694 | 648 | 593 | 458 | 407 | 309 | 475 |
| 26 | 589 | 572 | 620 | 574 | 556 | 618 | 638 | 559 | 489 | 359 | 280 | 375 |
| 27 | 573 | 574 | 613 | 595 | 629 | 616 | 579 | 524 | 515 | 488 | 409 | 410 |
| 28 | 637 | 675 | 648 | 668 | 669 | 684 | 662 | 625 | 588 | 552 | 525 | 624 |
| 29 | 747 | 720 | 773 | 680 | 634 | 705 | 664 | 585 | 464 | 521 | 447 | 443 |
| 30 | 755 | 779 | 700 | 772 | 782 | 764 | 718 | 606 | 636 | 623 | 553 | 521 |
| 31 | 625 | 589 | 566 | 690 | 686 | 706 | 688 | 609 | 549 | 456 | 494 | 561 |
| Mean. | 639 | 641 | 656 | 621 | 627 | 641 | 617 | 556 | 489 | 434 | 429 | 478 |

*zontal force, July,* 1889.

to .198000 dyne, give the absolute horizontal force in C. G. S. units.

.198000 +

| Day. | P. M. | | | | | | | | | | | | Mean. |
|---|---|---|---|---|---|---|---|---|---|---|---|---|---|
| | 1. | 2. | 3. | 4. | 5. | 6. | 7. | 8. | 9. | 10. | 11. | 12. | |
| 1 | 455 | 375 | 409 | 639 | 537 | 655 | 524 | 632 | 689 | 877 | 747 | 615 | 610 |
| 2 | 460 | 516 | 682 | 672 | ·607 | 579 | 618 | 632 | 642 | 671 | 845 | 742 | 630 |
| 3 | 657 | 760 | 823 | 818 | 767 | 546 | 627 | 632 | 628 | 619 | 615 | 634 | 618 |
| 4 | 577 | 676 | 714 | 757 | 682 | 668 | 636 | 665 | 690 | 695 | 701 | 729 | 670 |
| 5 | . . . | 677 | 602 | 476 | 547 | 665 | 703 | 698 | 732 | 698 | 700 | 644 | 647 |
| 6 | 611 | 550 | 499 | 593 | 627 | 650 | 618 | 586 | 497 | 676 | 611 | 564 | 625 |
| 7 | 653 | 705 | . . . | . . . | . . . | . . . | . . . | . . . | . . . | . . . | . . . | . . . | 680 |
| 8 | 380 | 676 | 743 | 753 | 758 | 650 | 675 | 693 | 680 | 652 | 664 | 659 | 648 |
| 9 | 668 | 696 | 725 | 702 | 689 | 665 | 676 | 680 | 658 | 677 | 673 | 659 | 641 |
| 10 | 757 | 790 | 753 | 749 | 787 | 721 | 657 | 685 | 691 | 691 | 663 | 668 | 651 |
| 11 | 484 | 705 | 730 | 683 | 646 | 670 | 619 | 628 | 629 | 648 | 649 | 668 | 626 |
| 12 | 550 | 621 | 683 | 706 | 670 | 576 | 581 | 656 | 662 | 657 | 631 | 664 | 567 |
| 13 | 697 | 809 | 820 | 810 | 717 | 600 | 657 | 671 | 682 | 677 | 678 | 664 | 626 |
| 14 | 546 | 739 | 627 | 815 | 788 | 708 | 624 | 587 | 475 | 494 | 452 | 485 | 562 |
| 15 | 602 | 649 | 627 | 627 | 647 | 703 | 685 | 685 | 700 | 672 | 706 | 603 | 604 |
| 16 | 453 | 581 | 592 | 668 | 650 | 646 | 647 | 658 | 659 | 664 | 656 | 1061 | 599 |
| 17 | 338 | 283 | 420 | 595 | 549 | 489 | 457 | 397 | 421 | 422 | 447 | 467 | 476 |
| 18 | 397 | 487 | 526 | 602 | 542 | 524 | 530 | 522 | 574 | 585 | 313 | 502 | 461 |
| 19 | 357 | 429 | 467 | 624 | 653 | 611 | 612 | 562 | 567 | 521 | 668 | 594 | 505 |
| 20 | 585 | 695 | 578 | 574 | 655 | 614 | 521 | 691 | 396 | 204 | 595 | 470 | 516 |
| 21 | 608 | 656 | 708 | 625 | 584 | 566 | 590 | 554 | 578 | 560 | 552 | 548 | 535 |
| 22 | 596 | 578 | 561 | 538 | 614 | 606 | 635 | 617 | 628 | 619 | 606 | 602 | 555 |
| 23 | 650 | 609 | 671 | 696 | 673 | 759 | 811 | 795 | 635 | 711 | 689 | 657 | 644 |
| 24 | 653 | 743 | 805 | 614 | 638 | 648 | 668 | 627 | 637 | 582 | 649 | 617 | 631 |
| 25 | 655 | 699 | 530 | 672 | 692 | 651 | 492 | 470 | 583 | 584 | 546 | 523 | 581 |
| 26 | 565 | 622 | 731 | 723 | 724 | 584 | 580 | 558 | 629 | 809 | 669 | 534 | 577 |
| 27 | 478 | 578 | 635 | 693 | 689 | 704 | 681 | 692 | 665 | 699 | 676 | 654 | 599 |
| 28 | 651 | 698 | 915 | 709 | 823 | 819 | 877 | 723 | 752 | 668 | 674 | 731 | .692 |
| 29 | 497 | 550 | 739 | 782 | 816 | 681 | 677 | 673 | 707 | 694 | 737 | 734 | 653 |
| 30 | 612 | 675 | 741 | 874 | 753 | 566 | 722 | 624 | 672 | 697 | 815 | 858 | 701 |
| 31 | 596 | 691 | 673 | 660 | 657 | 719 | 677 | 382 | 628 | 680 | 648 | 692 | 622 |
| Mean . | 560 | 630 | 658 | 682 | 673 | 641 | 636 | 623 | 626 | 637 | 643 | 641 | 605 |

TABLE VI—Continued—*Hori*

The figures given in the table are millionths of a dyne, which, add~d

.198000 +

| Day. | A. M. | | | | | | | | | | | |
|---|---|---|---|---|---|---|---|---|---|---|---|---|
| | 1. | 2. | 3. | 4. | 5. | 6. | 7. | 8. | 9. | 10. | 11. | 12. |
| 1 | 657 | 747 | 650 | 641 | 661 | 704 | 705 | 692 | 613 | 520 | 390 | 519 |
| 2 | 778 | 676 | 714 | 673 | 683 | 764 | 737 | 649 | 542 | 524 | 450 | 461 |
| 3 | 682 | 692 | 689 | 708 | 705 | 733 | 697 | 600 | 498 | 442 | 424 | 567 |
| 4 | 713 | 728 | 686 | 687 | 712 | 727 | 709 | 677 | 598 | 496 | 553 | 682 |
| 5 | 734 | 730 | 745 | 770 | 799 | 805 | 745 | 647 | 521 | 517 | 579 | 737 |
| 6 | 736 | 742 | 762 | 768 | 778 | 779 | 733 | 654 | 580 | 572 | 718 | 880 |
| 7 | 819 | 810 | 816 | 817 | 813 | 819 | 773 | 671 | 564 | 583 | 735 | 864 |
| 8 | 920 | 710 | 720 | 777 | 731 | 765 | 700 | 716 | 641 | 577 | 549 | 621 |
| 9 | 782 | 783 | 746 | 756 | 776 | 815 | 773 | 657 | 503 | 400 | 434 | 539 |
| 10 | 809 | 805 | 843 | 816 | 841 | 828 | 824 | 670 | 553 | 526 | 645 | 717 |
| 11 | 745 | 718 | 728 | 729 | 810 | 854 | 812 | 733 | 664 | 406 | 356 | 391 |
| 12 | 884 | 857 | 844 | 713 | 780 | 842 | 815 | 731 | 535 | 418 | 466 | 623 |
| 13 | 1028 | 648 | 724 | 734 | 806 | 821 | 662 | 640 | 570 | 360 | 427 | 588 |
| 14 | 767 | 782 | 708 | 695 | 663 | 706 | 712 | 675 | 563 | 499 | 518 | 586 |
| 15 | 783 | 860 | 785 | 749 | 914 | 936 | 907 | 913 | 716 | 576 | . . . | . . . |
| 16 | 781 | 810 | 882 | 855 | 828 | 810 | 736 | 708 | 625 | 560 | 617 | 680 |
| 17 | 840 | 846 | 832 | 828 | 834 | 810 | 835 | 751 | 615 | 705 | 795 | 885 |
| 18 | 823 | 820 | 801 | 806 | 789 | 822 | 766 | 617 | 518 | 557 | 614 | 746 |
| 19 | 793 | 813 | 808 | 805 | 820 | 820 | 755 | 700 | 643 | 621 | 692 | 711 |
| 20 | 955 | 768 | 777 | 694 | 883 | 690 | 743 | 828 | 734 | 590 | 558 | 487 |
| 21 | 845 | 738 | 780 | 767 | 759 | 754 | 769 | 733 | 751 | 691 | 711 | 692 |
| 22 | 753 | 773 | 749 | 722 | 728 | 770 | 729 | 631 | 509 | 468 | 567 | 718 |
| 23 | 845 | 813 | 832 | 805 | 820 | 825 | 783 | 728 | 620 | 489 | 476 | 641 |
| 24 | 838 | 829 | 853 | 826 | 836 | 827 | 781 | 711 | 631 | 632 | 657 | 741 |
| 25 | 883 | 874 | 799 | 828 | 858 | 891 | 826 | 573 | 451 | 616 | 848 | 1050 |
| 26 | 838 | 970 | 937 | 901 | 719 | 822 | 842 | 852 | 711 | 538 | 337 | 502 |
| 27 | 817 | 790 | 696 | 800 | 764 | 811 | 798 | 648 | 352 | 485 | 486 | 622 |
| 28 | 660 | 699 | 783 | 799 | 790 | 767 | 749 | 567 | 477 | 460 | 564 | 691 |
| 29 | 804 | 608 | 786 | 726 | 647 | 718 | 714 | 476 | 316 | 392 | 459 | 571 |
| 30 | 807 | 756 | 789 | 781 | 754 | 843 | 839 | 690 | 530 | 484 | 574 | 626 |
| 31 | 795 | 824 | 815 | 811 | 803 | 784 | 747 | 668 | 555 | 486 | 492 | 543 |
| Mean . | 805 | 775 | 777 | 765 | 778 | 796 | 765 | 684 | 571 | 522 | 560 | 656 |

*zontal force, August,* 1889.

to .198000 dyne, give the absolute horizontal force in C. G. S. units.

.198000 +

| Day. | 1. | 2. | 3. | 4. | 5. | 6. | 7. | 8. | 9. | 10. | 11. | 12. | Mean. |
|---|---|---|---|---|---|---|---|---|---|---|---|---|---|
| | | | | | | P. M. | | | | | | | |
| 1 | 788 | 643 | 747 | 584 | 684 | 718 | 677 | 631 | 731 | 591 | 658 | 688 | 652 |
| 2 | 547 | 675 | 850 | 855 | 847 | 726 | 703 | 643 | 672 | 711 | 712 | 784 | 682 |
| 3 | 723 | 752 | 805 | 783 | 774 | 705 | 692 | 697 | 736 | 732 | 700 | 698 | 676 |
| 4 | ... | 872 | 878 | 888 | 824 | 768 | 769 | 817 | 799 | 758 | 740 | 742 | 731 |
| 5 | 808 | 837 | 871 | 858 | 803 | 757 | 781 | 792 | 783 | 779 | 776 | 773 | 748 |
| 6 | 919 | 906 | 878 | 851 | 847 | 806 | 788 | 794 | 786 | 787 | 792 | 827 | 778 |
| 7 | 860 | 946 | 843 | 830 | 817 | 757 | 838 | 848 | 835 | 850 | 832 | 820 | 799 |
| 8 | 717 | 864 | 836 | 908 | 890 | 896 | 798 | 912 | 889 | 754 | 708 | 756 | 765 |
| 9 | 663 | 702 | 750 | 812 | 850 | 724 | 843 | 783 | 766 | 753 | 726 | 835 | 715 |
| 10 | 756 | 780 | 847 | 853 | 920 | 888 | 936 | 932 | 961 | 784 | 766 | 787 | 795 |
| 11 | 519 | 609 | 727 | 775 | 781 | 815 | 811 | 751 | 818 | 819 | 825 | 737 | 706 |
| 12 | 765 | 832 | 829 | 844 | 849 | 1005 | 917 | 946 | 835 | 709 | 785 | 834 | 777 |
| 13 | 556 | 933 | 742 | 658 | 654 | 749 | 572 | 615 | 738 | 749 | 759 | 789 | 688 |
| 14 | 672 | 814 | 918 | 910 | 864 | 780 | 762 | 810 | 802 | 784 | 837 | 820 | 735 |
| 15 | 735 | 887 | 855 | 790 | 880 | 848 | 802 | 695 | 720 | 646 | 586 | 714 | 786 |
| 16 | 714 | 800 | 839 | 797 | 709 | 654 | 749 | 754 | 798 | 808 | 814 | 834 | 757 |
| 17 | 970 | 943 | 962 | 892 | 809 | 785 | 777 | 783 | 787 | 798 | 775 | 808 | 819 |
| 18 | 789 | 771 | 818 | 810 | 806 | 745 | 802 | 761 | 789 | 795 | 815 | 801 | 758 |
| 19 | 712 | 845 | 859 | 860 | 837 | 851 | 862 | 834 | 849 | 878 | 860 | 869 | 796 |
| 20 | 639 | 682 | 677 | 453 | 510 | 552 | 732 | 803 | 644 | 772 | 782 | 805 | 698 |
| 21 | 721 | 718 | 863 | 832 | 809 | 739 | 740 | 773 | 750 | 742 | 743 | 771 | 758 |
| 22 | 784 | 767 | 818 | 824 | 773 | 773 | 788 | 794 | 813 | 800 | 815 | 815 | 737 |
| 23 | ... | 718 | 657 | 752 | 809 | 776 | 773 | 806 | 839 | 878 | 818 | 870 | 755 |
| 24 | 733 | 789 | 780 | 777 | 712 | 872 | 892 | 884 | 846 | 889 | 848 | 951 | 797 |
| 25 | 1051 | 1113 | 1051 | 968 | 805 | 875 | 820 | 915 | 816 | 864 | 837 | 865 | 853 |
| 26 | 709 | 743 | 696 | 669 | 557 | 698 | 695 | 677 | 710 | 678 | 707 | 698 | 717 |
| 27 | 736 | 892 | 897 | 818 | 654 | 706 | 730 | 783 | 778 | 794 | 799 | 842 | 729 |
| 28 | 790 | 890 | 801 | 750 | 761 | 732 | 522 | 593 | 777 | 712 | 840 | 652 | 701 |
| 29 | 718 | 733 | 790 | 861 | 740 | 825 | 796 | 780 | 765 | 710 | 697 | 805 | 685 |
| 30 | 627 | 713 | 802 | 831 | 776 | 738 | 763 | 820 | 820 | 816 | 794 | 766 | 739 |
| 31 | 629 | 776 | 888 | 871 | 876 | 853 | 868 | 827 | 798 | 828 | 819 | 815 | 757 |
| Mean. | 736 | 805 | 825 | 805 | 781 | 778 | 774 | 776 | 789 | 773 | 773 | 793 | 745 |

TABLE VI—Continued—*Hori*

The figures given in the table are millionths of a dyne, which, added

.198000 +

| Day. | A. M. | | | | | | | | | | | |
|------|------|------|------|------|------|------|------|------|------|------|------|------|
| | 1. | 2. | 3. | 4. | 5. | 6. | 7. | 8. | 9. | 10. | 11. | 12. |
| 1 | 849 | 850 | 850 | 851 | 867 | 881 | 863 | 723 | 558 | 447 | 391 | 462 |
| 2 | 819 | 882 | 830 | 826 | 785 | 907 | 861 | 674 | 575 | 520 | 540 | 624 |
| 3 | 892 | 883 | 874 | 889 | 900 | 900 | 816 | 695 | 582 | 606 | 645 | 687 |
| 4 | 820 | 788 | 811 | 789 | 794 | 790 | 758 | 613 | 566 | 623 | 681 | 704 |
| 5 | 864 | 865 | 865 | 847 | 853 | 839 | 802 | 685 | 685 | 686 | 687 | 711 |
| 6 | 843 | 858 | 872 | 906 | 917 | 888 | 880 | 811 | 749 | 694 | 672 | 686 |
| 7 | 832 | 833 | 833 | 839 | 849 | 849 | 808 | 635 | 569 | 579 | 627 | 717 |
| 8 | 801 | 727 | 652 | 732 | 762 | 738 | 833 | 726 | 623 | 544 | 498 | 615 |
| 9 | 442 | 678 | 368 | 444 | 535 | 431 | 380 | 475 | 395 | 298 | 209 | 172 |
| 10 | 637 | 511 | 722 | 718 | 560 | 597 | 495 | 425 | 294 | 271 | 239 | 230 |
| 11 | 466 | 819 | 556 | 566 | 577 | 661 | 582 | 438 | 236 | 227 | 266 | 430 |
| 12 | 656 | 653 | 667 | 607 | 749 | 664 | 651 | 539 | 375 | 296 | 311 | 386 |
| 13 | 584 | 665 | 721 | 628 | 705 | 709 | 654 | 495 | 349 | 350 | 403 | 539 |
| 14 | 708 | 661 | 667 | 659 | 674 | 675 | 610 | 498 | 443 | 406 | 416 | 510 |
| 15 | 701 | 682 | 730 | 708 | 756 | 757 | 725 | 674 | 581 | 493 | 475 | 542 |
| 16 | 703 | 727 | 742 | 780 | 777 | 759 | 783 | 705 | 654 | 537 | 543 | 633 |
| 17 | 734 | 734 | 753 | 778 | 751 | 742 | 725 | 622 | 567 | 483 | 484 | 565 |
| 18 | 722 | 745 | 765 | 790 | 748 | 754 | 741 | 620 | 536 | 518 | 482 | 539 |
| 19 | 663 | 672 | 673 | 722 | 732 | 723 | 654 | 551 | 529 | 492 | 460 | 499 |
| 20 | 670 | 689 | 708 | 709 | 720 | 711 | 661 | 554 | 479 | 396 | 477 | 543 |
| 21 | 742 | 742 | 753 | 759 | 736 | 751 | 696 | 560 | 514 | 492 | 507 | 569 |
| 22 | 763 | 707 | 619 | 714 | 649 | 617 | 886 | 788 | 578 | 179 | 064 | . . . |
| 23 | 474 | 531 | 348 | 519 | 543 | 530 | 569 | 424 | 307 | 332 | 051 | 395 |
| 24 | 458 | 519 | 482 | 572 | 583 | 598 | 622 | 511 | 375 | 367 | 387 | 557 |
| 25 | 676 | 667 | 649 | 692 | 538 | 605 | 662 | 503 | 340 | 397 | 455 | 498 |
| 26 | 556 | 556 | 656 | 647 | 662 | 640 | 585 | 553 | 516 | 475 | 480 | 491 |
| 27 | . . . | . . . | . . . | . . . | . . . | . . . | . . . | . . . | . . . | . . . | . . . | . . . |
| 28 | 617 | 645 | 651 | 647 | 709 | 640 | 598 | 496 | 426 | 413 | 443 | 481 |
| 29 | 699 | 690 | 709 | 710 | 693 | 679 | 629 | 540 | 476 | 406 | 468 | 540 |
| 30 | 720 | 706 | 674 | 694 | 695 | 649 | 669 | 618 | 553 | 479 | 452 | 491 |
| Mean . | 693 | 713 | 697 | 715 | 718 | 713 | 696 | 591 | 501 | 448 | 442 | 536 |

*zontal force, September,* 1889.

to .198000 dyne, give the absolute horizontal force in C. G. S. units.

.198000 +

| Day. | P. M. 1. | 2. | 3. | 4. | 5. | 6. | 7. | 8. | 9. | 10. | 11. | 12. | Mean. |
|---|---|---|---|---|---|---|---|---|---|---|---|---|---|
| 1 | . . . | 670 | 797 | 855 | 860 | 870 | 852 | 858 | 858 | 851 | 870 | 862 | 774 |
| 2 | 724 | 786 | 819 | 820 | 821 | 840 | 850 | 832 | 856 | 852 | 863 | 891 | 812 |
| 3 | 778 | 826 | 736 | 836 | 884 | 865 | 857 | 830 | 853 | 845 | 832 | 818 | 806 |
| 4 | 781 | 871 | 904 | 834 | 831 | 845 | 869 | 875 | 847 | 871 | 863 | 872 | 788 |
| 5 | 792 | 873 | 892 | 893 | 908 | 898 | 880 | 877 | 881 | 887 | 879 | 855 | 829 |
| 6 | . . . | 739 | 753 | 839 | 741 | 751 | 818 | 809 | 851 | 829 | 830 | 830 | 807 |
| 7 | 751 | 911 | 921 | 898 | 881 | 796 | 769 | 836 | 831 | 801 | 796 | 777 | 789 |
| 8 | 673 | 951 | 880 | 881 | 685 | 483 | 710 | 678 | 913 | 524 | 807 | 670 | 713 |
| 9 | 215 | 348 | 507 | 396 | 664 | 594 | 825 | 460 | 469 | 503 | 570 | 631 | 459 |
| 10 | 330 | 364 | 594 | 830 | 624 | 521 | 461 | 490 | 480 | 557 | 727 | 662 | 514 |
| 11 | 521 | 663 | 639 | 748 | 669 | 618 | 727 | 568 | 568 | 536 | 687 | 542 | 555 |
| 12 | . . . | 665 | 694 | 652 | 714 | 667 | 668 | 669 | 669 | 670 | 667 | 676 | 607 |
| 13 | 602 | 650 | 692 | 707 | 614 | 656 | 657 | 658 | 663 | 659 | 651 | 693 | 613 |
| 14 | 634 | 719 | 781 | 759 | 755 | 695 | 696 | 711 | 712 | 704 | 724 | 705 | 647 |
| 15 | 641 | 693 | 746 | 794 | 795 | 791 | 783 | 756 | 785 | 729 | 740 | 683 | 698 |
| 16 | 600 | 611 | 692 | 702 | 760 | 775 | 823 | 692 | 731 | 732 | 737 | 737 | 706 |
| 17 | 575 | 534 | 620 | 719 | 650 | 731 | 722 | 728 | 734 | 725 | 722 | 726 | 672 |
| 18 | 746 | 770 | 748 | 782 | 590 | 563 | 503 | 593 | 730 | 745 | 657 | 699 | 670 |
| 19 | 602 | 688 | 741 | 699 | 672 | 697 | 731 | 680 | 704 | 715 | 814 | 698 | 659 |
| 20 | 652 | 718 | 752 | 687 | 674 | 689 | 662 | 696 | 711 | 750 | 709 | 774 | 658 |
| 21 | 682 | 716 | 698 | 694 | 648 | 607 | 664 | 571 | 558 | 602 | 664 | 673 | 650 |
| 22 | 553 | 398 | 282 | 419 | 411 | 163 | 413 | 428 | 495 | 402 | 520 | 285 | 493 |
| 23 | 630 | 560 | 646 | 473 | 498 | 442 | 561 | 613 | 647 | 601 | 574 | 475 | 489 |
| 24 | 533 | 473 | 526 | 447 | 453 | 557 | 525 | 578 | 617 | 561 | 562 | 647 | 521 |
| 25 | 559 | 626 | 608 | 595 | 526 | 569 | 551 | 514 | 619 | 789 | 536 | 541 | 571 |
| 26 | . . . | 670 | 676 | . . . | . . . | . . . | . . . | . . . | . . . | . . . | . . . | . . . | 584 |
| 27 | 484 | 532 | 608 | 665 | 647 | 569 | 574 | 636 | 600 | 657 | 588 | 686 | 602 |
| 28 | 603 | 623 | 662 | 658 | 715 | 731 | 694 | 671 | 677 | 669 | 674 | 689 | 618 |
| 29 | 681 | 705 | 767 | 698 | 812 | 639 | 710 | 739 | 726 | 709 | 700 | 686 | 659 |
| 30 | 547 | 679 | 713 | 705 | 734 | 796 | 774 | 746 | 743 | 715 | 754 | 716 | 668 |
| Mean . | 611 | 670 | 707 | 713 | 697 | 670 | 701 | 682 | 708 | 696 | 714 | 697 | 655 |

TABLE VI—Continued—*Hori*

The figures given in the table are millionths of a dyne, which, added

.198000 +

| Day. | A. M. | | | | | | | | | | | |
|---|---|---|---|---|---|---|---|---|---|---|---|---|
| | 1. | 2. | 3. | 4. | 5. | 6. | 7. | 8. | 9. | 10. | 11. | 12. |
| 1 | 742 | 710 | 725 | 753 | 754 | 788 | 752 | 696 | 603 | 506 | 477 | 478 |
| 2 | 739 | 726 | 737 | 737 | 752 | 711 | 707 | 670 | 582 | 475 | 461 | 481 |
| 3 | 812 | 578 | 654 | 593 | 679 | 699 | 780 | 710 | 659 | 637 | 642 | 680 |
| 4 | 678 | 707 | 718 | 713 | 737 | 710 | 697 | 623 | 596 | 550 | 536 | 485 |
| 5 | 765 | 794 | 786 | 856 | 810 | 797 | 760 | 733 | 678 | 740 | 655 | 600 |
| 6 | 654 | 622 | 402 | 680 | 436 | 625 | 570 | 449 | 553 | 582 | 582 | 546 |
| 7 | 628 | 606 | 602 | 658 | 584 | 698 | 619 | 343 | 255 | 373 | 467 | 506 |
| 8 | 640 | 594 | 618 | 665 | 727 | 526 | 607 | 519 | 473 | 380 | 403 | 437 |
| 9 | 642 | 643 | 616 | 569 | 683 | 665 | 520 | 568 | 470 | 434 | 434 | 510 |
| 10 | 644 | 678 | 660 | 665 | 652 | 653 | 602 | 528 | 496 | 488 | 535 | 634 |
| 11 | 679 | 680 | 681 | 695 | 687 | 674 | 628 | 563 | 480 | 448 | 434 | 519 |
| 12 | 714 | 720 | 731 | 740 | 727 | 728 | 710 | 584 | 510 | 497 | 563 | 648 |
| 13 | 726 | 741 | 747 | 780 | 767 | 777 | 783 | 694 | 569 | 579 | 664 | 824 |
| 14 | 752 | 720 | 782 | 782 | 835 | 845 | 841 | 758 | 646 | 586 | 600 | 577 |
| 15 | 730 | 759 | 778 | 764 | 779 | 770 | 729 | 668 | 608 | 617 | 571 | 486 |
| 16 | 779 | 746 | 761 | 733 | 743 | 753 | 735 | 683 | 628 | 600 | 582 | 577 |
| 17 | 752 | 752 | 744 | 786 | 783 | 773 | 685 | 600 | 592 | 536 | 490 | 518 |
| 18 | 689 | 736 | 747 | 765 | 776 | 743 | 720 | 688 | 627 | 580 | 581 | 445 |
| 19 | 634 | 649 | 715 | 739 | 773 | 749 | 760 | 684 | 648 | 526 | 555 | 522 |
| 20 | 598 | 443 | 623 | 679 | 746 | 652 | 418 | 587 | 612 | 532 | 453 | 448 |
| 21 | 591 | 647 | 554 | 625 | 611 | 705 | 570 | 617 | 557 | 473 | 389 | 478 |
| 22 | 517 | 597 | 664 | 593 | 561 | 693 | 666 | 591 | 324 | 507 | 578 | 649 |
| 23 | 712 | 759 | 704 | 718 | 719 | 794 | 748 | 668 | 683 | 594 | 619 | 605 |
| 24 | 705 | 686 | 701 | 701 | 716 | 716 | 708 | 623 | 577 | 549 | 597 | 630 |
| 25 | 669 | 664 | 688 | 684 | 689 | 713 | 704 | 554 | 518 | 433 | 434 | 434 |
| 26 | 722 | 745 | 737 | 760 | 799 | 780 | 772 | 692 | 543 | 505 | 562 | 600 |
| 27 | 714 | 728 | 748 | 753 | 754 | 782 | 750 | 651 | 558 | 436 | 442 | 465 |
| 28 | 717 | 712 | 718 | 798 | 780 | 780 | 748 | 701 | 528 | 467 | 501 | 581 |
| 29 | 699 | 685 | 705 | 700 | 706 | 711 | 684 | 608 | 530 | 497 | 512 | 531 |
| 30 | 725 | 725 | 773 | 726 | 736 | 736 | 742 | 699 | 672 | 597 | 575 | 542 |
| 31 | 750 | 731 | 746 | 765 | 780 | 785 | 757 | 739 | 683 | 627 | 595 | 703 |
| Mean . | 694 | 687 | 696 | 715 | 719 | 727 | 693 | 629 | 563 | 577 | 532 | 553 |

*zontal force, October,* 1889.

to .198000 dyne, give the absolute horizontal force in C. G. S. units.

.198000 +

| Day. | P. M. | | | | | | | | | | | | Mean. |
| --- | --- | --- | --- | --- | --- | --- | --- | --- | --- | --- | --- | --- | --- |
| | 1. | 2. | 3. | 4. | 5. | 6. | 7. | 8. | 9. | 10. | 11. | 12. | |
| 1 | 526 | 621 | 641 | 713 | 737 | 743 | 733 | 692 | 726 | 722 | 733 | 734 | 684 |
| 2 | 613 | 755 | 808 | 799 | 805 | 820 | 811 | 835 | 822 | 842 | 834 | 774 | 721 |
| 3 | 770 | 720 | 777 | 726 | 629 | 700 | 691 | 697 | 688 | 694 | 685 | 686 | 691 |
| 4 | 571 | 614 | 676 | 705 | 772 | 820 | 796 | 793 | 780 | 771 | 749 | 759 | 690 |
| 5 | 648 | 672 | 692 | 430 | 304 | 394 | 239 | 170 | 401 | 548 | 478 | 226 | 591 |
| 6 | 486 | 618 | 586 | 531 | 372 | 533 | 636 | 656 | 629 | 583 | 598 | 514 | 560 |
| 7 | 446 | 517 | 542 | 608 | 562 | 535 | 587 | 560 | 659 | 524 | 704 | 625 | 550 |
| 8 | 514 | 623 | 483 | 662 | 673 | 674 | 702 | 646 | 539 | 611 | 541 | 618 | 578 |
| 9 | 473 | 578 | 527 | 556 | 595 | 535 | 530 | 653 | 650 | 632 | 638 | 643 | 574 |
| 10 | 734 | 707 | 717 | 638 | 625 | 669 | 669 | 674 | 685 | 681 | 677 | 688 | 642 |
| 11 | 591 | 690 | 715 | 744 | 731 | 727 | 741 | 742 | 715 | 711 | 698 | 695 | 653 |
| 12 | 696 | 763 | 792 | 807 | 775 | 767 | 767 | 730 | 717 | 775 | 729 | 725 | 705 |
| 13 | 769 | 648 | 804 | 786 | 707 | 694 | 718 | 714 | 691 | 753 | 726 | 765 | 726 |
| 14 | 621 | 561 | 632 | 751 | 794 | 781 | 767 | 740 | 736 | 756 | 719 | 781 | 723 |
| 15 | 572 | 516 | 601 | 737 | 428 | 635 | 683 | 678 | 698 | 750 | 676 | 746 | 666 |
| 16 | 635 | 677 | 697 | 744 | 768 | 712 | 755 | 760 | 723 | 719 | 743 | 752 | 708 |
| 17 | 641 | 683 | 774 | 792 | 793 | 770 | 780 | 780 | 744 | 762 | 684 | 712 | 705 |
| 18 | 511 | 478 | 568 | 479 | 245 | 447 | 533 | 542 | 473 | 435 | 525 | 591 | 576 |
| 19 | 715 | 729 | 739 | 640 | 576 | 604 | 426 | 633 | 484 | 596 | 682 | 607 | 641 |
| 20 | 506 | 577 | 526 | 624 | 588 | 602 | 499 | 514 | 590 | 576 | 694 | 741 | 576 |
| 21 | 677 | 550 | 584 | 631 | 505 | 594 | 618 | 591 | 648 | 709 | 672 | 672 | 595 |
| 22 | 664 | 655 | 600 | 562 | 591 | 610 | 653 | 653 | 687 | 710 | 711 | 688 | 614 |
| 23 | 614 | 656 | 657 | 704 | 738 | 701 | 613 | 565 | 576 | 543 | 586 | 638 | 663 |
| 24 | 672 | 738 | 711 | 649 | 603 | 674 | 656 | 642 | 643 | 634 | 639 | 668 | 660 |
| 25 | 505 | 552 | 619 | 685 | 714 | 742 | 767 | 734 | 707 | 702 | 680 | 717 | 638 |
| 26 | 587 | 686 | 743 | 743 | 758 | 749 | 768 | 745 | 760 | 741 | 719 | 733 | 706 |
| 27 | 551 | 650 | 721 | 768 | 793 | 717 | 770 | 775 | 766 | 738 | 706 | 660 | 683 |
| 28 | 665 | 712 | 694 | 685 | 634 | 691 | 734 | 762 | 697 | 660 | 623 | 637 | 676 |
| 29 | 596 | 677 | 771 | 814 | 815 | 829 | 736 | 684 | 652 | 784 | 691 | 686 | 679 |
| 30 | 590 | 627 | 675 | 722 | 747 | 751 | 729 | 715 | 744 | 716 | 764 | 740 | 699 |
| 31 | 727 | 723 | 691 | 653 | 682 | 659 | 730 | 763 | 746 | 755 | 761 | 798 | 723 |
| Mean . | 609 | 644 | 670 | 680 | 647 | 674 | 672 | 672 | 670 | 682 | 680 | 678 | 655 |

TABLE VI—Continued—*Hori*

The figures given in the table are millionths of a dyne, which, added

.198000 +

| Day. | A. M. | | | | | | | | | | | |
|------|-----|------|-----|-----|-----|-----|-----|-----|-----|--------|--------|-----|
| | 1. | 2. | 3. | 4. | 5. | 6. | 7. | 8. | 9. | 10. | 11. | 12. |
| 1 | 750 | 1064 | 572 | 722 | 667 | 714 | 729 | 466 | 345 | (—)135 | (—)087 | 247 |
| 2 | 399 | 404 | 536 | 452 | 528 | 514 | 454 | 547 | 342 | 323 | 573 | 625 |
| 3 | 557 | 660 | 567 | 473 | 676 | 526 | 639 | 273 | 523 | 514 | 458 | 411 |
| 4 | 647 | 638 | 587 | 597 | 663 | 640 | 660 | 561 | 459 | 520 | 540 | 549 |
| 5 | 668 | 635 | 612 | 631 | 674 | 552 | 694 | 708 | 521 | 470 | 546 | 518 |
| 6 | 599 | 637 | 595 | 666 | 685 | 667 | 672 | 625 | 579 | 561 | 613 | 660 |
| 7 | 676 | 671 | 663 | 691 | 701 | 724 | 702 | 693 | 604 | 567 | 601 | 610 |
| 8 | 636 | 645 | 637 | 675 | 732 | 718 | 724 | 663 | 602 | 565 | 603 | 636 |
| 9 | 736 | 736 | 751 | 770 | 776 | 766 | 753 | 674 | 623 | 590 | 577 | 582 |
| 10 | 686 | 644 | 664 | 701 | 735 | 707 | 699 | 637 | 582 | 526 | 536 | 555 |
| 11 | 646 | 688 | 708 | 745 | 741 | 765 | 766 | 752 | 715 | 697 | 669 | 613 |
| 12 | 746 | 746 | 723 | 756 | 752 | 799 | 735 | 669 | 599 | 599 | 582 | 600 |
| 13 | 744 | 739 | 763 | 749 | 779 | 779 | 761 | 690 | 635 | 616 | 589 | 580 |
| 14 | 783 | 764 | 793 | 760 | 804 | 790 | 800 | 791 | 763 | 740 | 736 | 666 |
| 15 | 766 | 770 | 784 | 822 | 823 | 809 | 809 | 818 | 763 | 641 | 580 | . . . |
| 16 | 650 | 828 | 631 | 668 | 716 | 745 | 754 | 693 | 637 | 581 | 581 | 558 |
| 17 | 721 | 726 | 731 | 768 | 859 | 732 | 873 | 797 | 479 | 446 | 517 | 451 |
| 18 | 789 | 507 | 601 | 639 | 649 | 687 | 696 | 659 | 622 | 500 | 336 | 495 |
| 19 | 617 | 631 | 631 | 673 | 679 | 707 | 702 | 726 | 708 | 675 | 633 | 586 |
| 20 | 656 | 670 | 675 | 693 | 708 | 722 | 704 | 690 | 644 | 606 | 573 | 526 |
| 21 | 616 | 644 | 677 | 705 | 715 | 711 | 706 | 715 | 632 | 571 | . . . | . . . |
| 22 | 655 | 655 | 664 | 678 | 670 | 698 | 674 | 759 | 713 | 633 | 530 | 483 |
| 23 | 661 | 684 | 679 | 712 | 723 | 732 | 723 | 718 | 667 | 630 | 578 | 531 |
| 24 | 757 | 776 | 743 | 715 | 716 | 744 | 810 | 792 | 773 | 764 | 698 | 688 |
| 25 | 683 | 650 | 678 | 716 | 703 | 750 | 740 | 816 | 760 | 558 | 591 | 633 |
| 26 | 684 | 727 | 736 | 764 | 723 | 695 | 808 | 709 | 559 | 310 | 240 | 136 |
| 27 | 315 | 522 | 489 | 503 | 533 | 575 | 598 | 622 | 628 | 595 | 510 | 468 |
| 28 | 641 | 617 | 660 | 674 | 703 | 788 | 844 | 900 | 775 | 629 | 511 | 027 |
| 29 | 586 | 497 | 539 | 595 | 610 | 676 | 568 | 606 | 522 | 456 | 221 | 343 |
| 30 | 583 | 616 | 640 | 659 | 589 | 693 | 580 | 697 | 642 | 538 | 369 | 416 |
| Mean . | 655 | 673 | 658 | 679 | 701 | 704 | 709 | 683 | 614 | 540 | 517 | 510 |

*zontal force, November,* 1889.

to .198000 dyne, give the absolute horizontal force in C. G. S. units.

.198000 +

| Day. | P. M. | | | | | | | | | | | | Mean. |
|---|---|---|---|---|---|---|---|---|---|---|---|---|---|
| | 1. | 2. | 3. | 4. | 5. | 6. | 7. | 8. | 9. | 10. | 11. | 12. | |
| 1 | 201 | 403 | 329 | 447 | 114 | 208 | (—)181 | 091 | 139 | 200 | 418 | 554 | 378 |
| 2 | 569 | 513 | 411 | 523 | 543 | 520 | 610 | 521 | 381 | 475 | 546 | 565 | 495 |
| 3 | 421 | 477 | 563 | 614 | 620 | 639 | 649 | 635 | 692 | 725 | 585 | 628 | 564 |
| 4 | 610 | 601 | 602 | 630 | 654 | 617 | 557 | 618 | 525 | 638 | 587 | 728 | 601 |
| 5 | 533 | 523 | 562 | 604 | 568 | 619 | 559 | 409 | 419 | 523 | 547 | 632 | 572 |
| 6 | 704 | 713 | 690 | 714 | 701 | 715 | 725 | 711 | 722 | 717 | 803 | 676 | 673 |
| 7 | 625 | 691 | 720 | 669 | 670 | 698 | 713 | 708 | 667 | 648 | 640 | 630 | 666 |
| 8 | 688 | 744 | 689 | 708 | 751 | 756 | 747 | 729 | 701 | 697 | 702 | 735 | 687 |
| 9 | 676 | 755 | 803 | 850 | 837 | 804 | 542 | 552 | 665 | 637 | 681 | 751 | 704 |
| 10 | 603 | 636 | 707 | 693 | 769 | 732 | 648 | 709 | 720 | 663 | 669 | 664 | 662 |
| 11 | 708 | 722 | 793 | 770 | 771 | 794 | 781 | 725 | 745 | 707 | 713 | 713 | 727 |
| 12 | 667 | 714 | 767 | 800 | 768 | 763 | 731 | 713 | 685 | 690 | 729 | 724 | 711 |
| 13 | 584 | 669 | 717 | 759 | 784 | 798 | 789 | 803 | 790 | 790 | 796 | 782 | 729 |
| 14 | 670 | 685 | 761 | 803 | 818 | 823 | 767 | 711 | 750 | 773 | 737 | 760 | 760 |
| 15 | 609 | 684 | 656 | 717 | 671 | 601 | 652 | 633 | 649 | 691 | 672 | 691 | 709 |
| 16 | 629 | 653 | 638 | 638 | 733 | 710 | 705 | 639 | 669 | 678 | 706 | 702 | 664 |
| 17 | 400 | 532 | 438 | 377 | 481 | 547 | 589 | 448 | 552 | 501 | 585 | 520 | 586 |
| 18 | 497 | 525 | 553 | 633 | 648 | 690 | 714 | 648 | 621 | 592 | 564 | 602 | 603 |
| 19 | 610 | 643 | 700 | 676 | 682 | 677 | 663 | 658 | 655 | 692 | 645 | 650 | 663 |
| 20 | 579 | 631 | 663 | 748 | 721 | 702 | 702 | 702 | 689 | 670 | 604 | 595 | 661 |
| 21 | 619 | 685 | 727 | 760 | 784 | 737 | 869 | 704 | 569 | 602 | 569 | 658 | 681 |
| 22 | 564 | 644 | 738 | 780 | 757 | 729 | 734 | 706 | 740 | 664 | 636 | 622 | 672 |
| 23 | 659 | 720 | 725 | 753 | 735 | 735 | 749 | 726 | 722 | 722 | 750 | 727 | 697 |
| 24 | 802 | 821 | 901 | 534 | 742 | 700 | 728 | 672 | 536 | 555 | 654 | 701 | 726 |
| 25 | 757 | 785 | 832 | 841 | 743 | 743 | 673 | 692 | 679 | 721 | 693 | 693 | 714 |
| 26 | 213 | 720 | 105 | 227 | 279 | 279 | 110 | 026 | 262 | 229 | 384 | 586 | 438 |
| 27 | 328 | 332 | 356 | 445 | 446 | 343 | 390 | 460 | 569 | 513 | 640 | 491 | 491 |
| 28 | (—)005 | 409 | 489 | 559 | 382 | 372 | 344 | 452 | 449 | 458 | 505 | 458 | 527 |
| 29 | 504 | 584 | 641 | 683 | 679 | 482 | 566 | 613 | 610 | 652 | 624 | 628 | 554 |
| 30 | 380 | 483 | 549 | 619 | 672 | 667 | 625 | 616 | 579 | 645 | 668 | 673 | 592 |
| Mean. | 547 | 623 | 627 | 649 | 651 | 640 | 612 | 601 | 618 | 629 | 635 | 655 | 630 |

TABLE VI—Continued—*Hori*

The figures given in the table are millionths of a dyne, which, added

. 198000 +

| Day. | A. M. | | | | | | | | | | | |
|---|---|---|---|---|---|---|---|---|---|---|---|---|
| | 1. | 2. | 3. | 4. | 5. | 6. | 7. | 8. | 9. | 10. | 11. | 12. |
| 1 | 663 | 677 | 705 | 649 | 711 | 664 | 706 | 678 | 642 | 656 | 552 | 496 |
| 2 | 650 | 655 | 678 | 674 | 698 | 689 | 722 | 689 | 699 | 586 | 460 | 445 |
| 3 | 662 | 714 | 695 | 709 | 677 | 705 | 733 | 724 | 763 | 730 | 655 | 481 |
| 4 | 701 | 663 | 677 | 677 | 683 | 702 | 721 | 754 | 750 | 693 | 689 | 661 |
| 5 | 665 | 716 | 697 | 707 | 717 | 736 | 755 | 736 | 714 | 695 | 667 | 695 |
| 6 | 609 | 666 | 619 | 722 | 794 | 812 | 841 | 808 | 724 | 607 | 734 | 673 |
| 7 | 626 | 626 | 659 | 668 | 721 | 768 | 716 | 744 | 722 | 694 | 647 | 557 |
| 8 | 669 | 665 | 726 | 688 | 699 | 670 | 661 | 642 | 568 | 516 | 474 | 521 |
| 9 | 728 | 738 | 723 | 709 | 743 | 743 | 743 | 715 | 646 | 561 | 490 | 519 |
| 10 | 701 | 725 | 734 | 748 | 749 | 754 | 740 | 735 | 713 | 652 | 558 | 558 |
| 11 | 717 | 740 | 726 | 740 | 746 | 746 | 736 | 732 | 737 | 705 | 596 | 573 |
| 12 | 760 | 760 | 765 | 770 | 799 | 799 | 813 | 846 | 823 | 767 | 664 | 631 |
| 13 | 767 | 744 | 744 | 800 | 838 | 810 | 833 | 871 | 881 | 778 | 745 | 670 |
| 14 | 679 | 745 | 590 | 675 | 685 | 704 | 746 | 798 | 789 | 785 | 724 | 672 |
| 15 | 662 | 652 | 742 | 695 | 719 | 710 | 747 | 743 | 758 | 697 | 612 | 560 |
| 16 | 716 | 725 | 749 | 786 | 815 | 881 | 830 | 806 | 779 | 704 | 581 | 614 |
| 17 | 628 | 647 | 679 | 726 | 657 | 746 | 774 | 751 | 714 | 639 | 540 | 498 |
| 18 | 728 | 709 | 714 | 732 | 748 | 795 | 785 | 795 | 796 | 749 | 617 | 565 |
| 19 | 669 | 655 | 683 | 744 | 759 | 778 | 797 | 787 | 798 | 690 | 549 | 544 |
| 20 | 731 | 694 | 713 | 769 | 793 | 765 | 812 | 681 | 719 | 715 | 588 | 574 |
| 21 | 672 | 723 | 719 | 747 | 724 | 752 | 795 | 762 | 763 | 706 | 556 | 476 |
| 22 | 711 | 664 | 688 | 702 | 722 | 745 | 722 | 745 | 709 | 671 | 572 | 493 |
| 23 | 647 | 699 | 680 | 689 | 615 | 667 | 695 | 718 | 743 | 696 | 583 | 593 |
| 24 | 719 | 738 | 733 | 695 | 715 | 720 | 748 | 724 | 584 | 509 | 462 | 514 |
| 25 | 716 | 712 | 707 | 773 | 764 | 764 | 797 | 793 | 732 | 629 | 554 | 577 |
| 26 | 746 | 746 | 732 | 765 | 770 | 775 | 761 | 775 | 710 | 673 | 588 | 546 |
| 27 | 465 | 272 | 366 | 479 | 438 | 555 | 541 | 574 | 594 | 566 | 575 | 481 |
| 28 | 665 | 660 | 702 | 698 | 718 | 733 | 718 | 746 | 700 | 418 | 150 | 437 |
| 29 | 586 | 671 | 694 | 666 | 672 | 657 | 709 | 662 | 673 | 537 | 424 | 358 |
| 30 | 592 | 606 | 630 | 672 | 687 | 612 | 706 | 687 | 594 | 477 | 406 | 453 |
| 31 | ... | ... | ... | ... | ... | ... | ... | ... | ... | ... | ... | ... |
| Mean . | 675 | 680 | 689 | 709 | 719 | 732 | 747 | 741 | 718 | 650 | 567 | 548 |

*zontal force, December,* 1889.

to .198000 dyne, give the absolute horizontal force in C. G. S. units.

.198000 +

| Day. | P. M. | | | | | | | | | | | | Mean. |
|---|---|---|---|---|---|---|---|---|---|---|---|---|---|
| | 1. | 2. | 3. | 4. | 5. | 6. | 7. | 8. | 9. | 10. | 11. | 12. | |
| 1 | 563 | 525 | 605 | 600 | 620 | 639 | 644 | 639 | 649 | 663 | 621 | 663 | 635 |
| 2 | 517 | 531 | 606 | 686 | 673 | 607 | 626 | 678 | 679 | 669 | 679 | 669 | 635 |
| 3 | 547 | 613 | 656 | 656 | 624 | 642 | 694 | 689 | 690 | 695 | 667 | 667 | 670 |
| 4 | 652 | 643 | 680 | 657 | 724 | 724 | 724 | 714 | 692 | 706 | 692 | 682 | 694 |
| 5 | 700 | 738 | 752 | 710 | 650 | 673 | 730 | 753 | 773 | 726 | 679 | 665 | 710 |
| 6 | 584 | 518 | 683 | .702 | 646 | 646 | 604 | 646 | 530 | 610 | 591 | 605 | 666 |
| 7 | 582 | 666 | 666 | 615 | 620 | 672 | 672 | 531 | 701 | 640 | 612 | 730 | 661 |
| 8 | 522 | 640 | 668 | 691 | 735 | 758 | 716 | 673 | 632 | 642 | 712 | 684 | 649 |
| 9 | 567 | 600 | 703 | 675 | 732 | 648 | 718 | 723 | 724 | 714 | 714 | 696 | 678 |
| 10 | 606 | 685 | 742 | 732 | 733 | 719 | 743 | 748 | 678 | 711 | 725 | 716 | 704 |
| 11 | 597 | 649 | 710 | 800 | 829 | 805 | 786 | 768 | 745 | 759 | 750 | 764 | 727 |
| 12 | 674 | 693 | 735 | 754 | 755 | 713 | 727 | 708 | 704 | 746 | 756 | 742 | 746 |
| 13 | 694 | 746 | 769 | 647 | 756 | 794 | 747 | 775 | 720 | 687 | 602 | 494 | 746 |
| 14 | 659 | 649 | 631 | 635 | 655 | 646 | 613 | 655 | 665 | 670 | 637 | 656 | 682 |
| 15 | 561 | 627 | 693 | 721 | 699 | 689 | 666 | 689 | 690 | 686 | 690 | 700 | 684 |
| 16 | 662 | 639 | 709 | 799 | 753 | 541 | 485 | 654 | 683 | 688 | 646 | 674 | 705 |
| 17 | 570 | 598 | 654 | 701 | 740 | 698 | 735 | 707 | 694 | 699 | 694 | 694 | 674 |
| 18 | 604 | 703 | 773 | 825 | 807 | 765 | 755 | 718 | 705 | 747 | 733 | 719 | 733 |
| 19 | 625 | 691 | 771 | 846 | 786 | 725 | 701 | 692 | 740 | 782 | 796 | 740 | 723 |
| 20 | 640 | 650 | 650 | 767 | 773 | 637 | 670 | 599 | 718 | 732 | 685 | 680 | 698 |
| 21 | 477 | 505 | 505 | 538 | 704 | 723 | 732 | 727 | 686 | 531 | 521 | 695 | 656 |
| 22 | 578 | 649 | 611 | 639 | 706 | 692 | 701 | 697 | 712 | 646 | 618 | 684 | 670 |
| 23 | 617 | 594 | 688 | 720 | 726 | 747 | 721 | 707 | 671 | 600 | 638 | 647 | 671 |
| 24 | 618 | 731 | 797 | 769 | 751 | 746 | 704 | 671 | 799 | 705 | 696 | 700 | 690 |
| 25 | 611 | 701 | 752 | 823 | 791 | 753 | 730 | 716 | 717 | 726 | 712 | 717 | 719 |
| 26 | 641 | 707 | 819 | 791 | 764 | 787 | 698 | 463 | 464 | 544 | 582 | 417 | 678 |
| 27 | 576 | 694 | 788 | 835 | 812 | 812 | 784 | 770 | 747 | 729 | 639 | 630 | 613 |
| 28 | 541 | 616 | 611 | 658 | 744 | 674 | 598 | 674 | 637 | 689 | 670 | 618 | 628 |
| 29 | 415 | 481 | 594 | 608 | 741 | 698 | 661 | 637 | 605 | 605 | 577 | 615 | 606 |
| 30 | 327 | 412 | 506 | 600 | 615 | 606 | 629 | 629 | 635 | 625 | 635 | 621 | 582 |
| 31 | ... | ... | ... | | | | | | | | | | ... |
| Mean. | 584 | 630 | 684 | 707 | 722 | 699 | 690 | 682 | 683 | 679 | 665 | 666 | 678 |

TABLE VII.—*Ver*

The figures given in the table are millionths of a dyne, which,

.581000 +

| Day. | A. M. | | | | | | | | | | | |
|---|---|---|---|---|---|---|---|---|---|---|---|---|
| | 1. | 2. | 3. | 4. | 5. | 6. | 7. | 8. | 9. | 10. | 11. | 12. |
| 1 | . . . | . . . | . . . | . . . | . . . | . . . | . . . | . . . | . . . | . . . | . . . | . . . |
| 2 | . . . | . . . | . . . | . . . | . . . | . . . | . . . | . . . | . . . | . . . | . . . | . . . |
| 3 | . . . | . . . | . . . | . . . | . . . | . . . | . . . | . . . | . . . | . . . | . . . | . . . |
| 4 | . . . | . . . | . . . | . . . | . . . | . . . | . . . | . . . | . . . | . . . | . . . | . . . |
| 5 | . . . | . . . | . . . | . . . | . . . | . . . | . . . | . . . | . . . | . . . | . . . | . . . |
| 6 | . . . | . . . | . . . | . . . | . . . | . . . | . . . | . . . | . . . | . . . | . . . | . . . |
| 7 | . . . | . . . | . . . | . . . | . . . | . . . | . . . | . . . | . . . | . . . | . . . | . . . |
| 8 | . . . | . . . | . . . | . . . | . . . | . . . | . . . | . . . | . . . | . . . | . . . | . . . |
| 9 | . . . | . . . | . . . | . . . | . . . | . . . | . . . | . . . | . . . | . . . | . . . | . . . |
| 10 | . . . | . . . | . . . | . . . | . . . | . . . | . . . | . . . | . . . | . . . | . . . | . . . |
| 11 | . . . | . . . | . . . | . . . | . . . | . . . | . . . | . . . | . . . | . . . | . . . | . . . |
| 12 | . . . | . . . | . . . | . . . | . . . | . . . | . . . | . . . | . . . | . . . | . . . | . . . |
| 13 | . . . | . . . | . . . | . . . | . . . | . . . | . . . | . . . | . . . | . . . | . . . | . . . |
| 14 | . . . | . . . | . . . | . . . | . . . | . . . | . . . | . . . | . . . | . . . | . . . | . . . |
| 15 | . . . | . . . | . . . | . . . | . . . | . . . | . . . | . . . | . . . | . . . | . . . | . . . |
| 16 | . . . | . . . | . . . | . . . | . . . | . . . | . . . | . . . | . . . | . . . | . . . | . . . |
| 17 | . . . | . . . | . . . | . . . | . . . | . . . | . . . | . . . | . . . | . . . | . . . | . . . |
| 18 | . . . | . . . | . . . | . . . | . . . | . . . | . . . | . . . | . . . | . . . | . . . | . . . |
| 19 | . . . | . . . | . . . | . . . | . . . | . . . | . . . | . . . | . . . | . . . | . . . | . . . |
| 20 | . . . | . . . | . . . | . . . | . . . | . . . | . . . | . . . | . . . | . . . | . . . | . . . |
| 21 | . . . | . . . | . . . | . . . | . . . | . . . | . . . | . . . | . . . | . . . | . . . | . . . |
| 22 | . . . | . . . | . . . | . . . | . . . | . . . | . . . | . . . | . . . | . . . | . . . | . . . |
| 23 | . . . | . . . | . . . | . . . | . . . | . . . | . . . | . . . | . . . | . . . | . . . | . . . |
| 24 | . . . | . . . | . . . | . . . | . . . | . . . | . . . | . . . | . . . | . . . | . . . | . . . |
| 25 | . . . | . . . | . . . | . . . | . . . | . . . | . . . | . . . | . . . | . . . | . . . | . . . |
| 26 | 916 | 916 | 916 | 916 | 916 | 916 | 911 | 887 | 829 | 796 | 748 | 748 |
| 27 | 892 | 892 | 896 | 896 | 896 | 896 | 896 | 896 | 844 | 820 | 772 | . . . |
| 28 | 896 | 895 | 879 | 891 | 881 | 889 | 911 | 895 | 889 | 869 | 886 | 875 |
| 29 | 869 | 873 | 871 | 864 | 858 | 862 | 860 | 853 | 843 | 818 | 782 | 790 |
| 30 | 630 | 620 | 618 | 621 | 615 | 633 | 617 | 610 | 590 | 593 | 563 | 561 |
| 31 | 627 | 607 | 619 | 622 | 616 | 620 | 618 | 650 | 653 | 686 | 636 | 624 |
| Mean . | 805 | 801 | 800 | 802 | 797 | 803 | 802 | 799 | 774 | 764 | 731 | 718 |

*tical force, January,* 1889.

added to .581000 dyne, give the vertical force in C. G. S. units.

.581000 +

| Day. | P. M. | | | | | | | | | | | | Mean. |
|---|---|---|---|---|---|---|---|---|---|---|---|---|---|
| | 1. | 2. | 3. | 4. | 5. | 6. | 7. | 8. | 9. | 10. | 11. | 12. | |
| 1 | . . . | . . . | . . . | . . . | . . . | . . . | . . . | . . . | . . . | . . . | . . . | . . . | . . . |
| 2 | . . . | . . . | . . . | . . . | . . . | . . . | . . . | . . . | . . . | . . . | . . . | . . . | . . . |
| 3 | . . . | . . . | . . . | . . . | . . . | . . . | . . . | . . . | . . . | . . . | . . . | . . . | . . . |
| 4 | . . . | . . . | . . . | . . . | . . . | . . . | . . . | . . . | . . . | . . . | . . . | . . . | . . . |
| 5 | . . . | . . . | . . . | . . . | . . . | . . . | . . . | . . . | . . . | . . . | . . . | . . . | . . . |
| 6 | . . . | . . . | . . . | . . . | . . . | . . . | . . . | . . . | . . . | . . . | . . . | . . . | . . . |
| 7 | . . . | . . . | . . . | . . . | . . . | . . . | . . . | . . . | . . . | . . . | . . . | . . . | . . . |
| 8 | . . . | . . . | . . . | . . . | . . . | . . . | . . . | . . . | . . . | . . . | . . . | . . . | . . . |
| 9 | . . . | . . . | . . . | . . . | . . . | . . . | . . . | . . . | . . . | . . . | . . . | . . . | . . . |
| 10 | . . . | . . . | . . . | . . . | . . . | . . . | . . . | . . . | . . . | . . . | . . . | . . . | . . . |
| 11 | . . . | . . . | . . . | . . . | . . . | . . . | . . . | . . . | . . . | . . . | . . . | . . . | . . . |
| 12 | . . . | . . . | . . . | . . . | . . . | . . . | . . . | . . . | . . . | . . . | . . . | . . . | . . . |
| 13 | . . . | . . . | . . . | . . . | . . . | . . . | . . . | . . . | . . . | . . . | . . . | . . . | . . . |
| 14 | . . . | . . . | . . . | . . . | . . . | . . . | . . . | . . . | . . . | . . . | . . . | . . . | . . . |
| 15 | . . . | . . . | . . . | . . . | . . . | . . . | . . . | . . . | . . . | . . . | . . . | . . . | . . . |
| 16 | . . . | . . . | . . . | . . . | . . . | . . . | . . . | . . . | . . . | . . . | . . . | . . . | . . . |
| 17 | . . . | . . . | . . . | . . . | . . . | . . . | . . . | . . . | . . . | . . . | . . . | . . . | . . . |
| 18 | . . . | . . . | . . . | . . . | . . . | . . . | . . . | . . . | . . . | . . . | . . . | . . . | . . . |
| 19 | . . . | . . . | . . . | . . . | . . . | . . . | . . . | . . . | . . . | . . . | . . . | . . . | . . . |
| 20 | . . . | . . . | . . . | . . . | . . . | . . . | . . . | . . . | . . . | . . . | . . . | . . . | . . . |
| 21 | . . . | . . . | . . . | . . . | . . . | . . . | . . . | . . . | . . . | . . . | . . . | . . . | . . . |
| 22 | . . . | . . . | . . . | . . . | . . . | . . . | . . . | . . . | . . . | . . . | . . . | . . . | . . . |
| 23 | . . . | . . . | . . . | . . . | . . . | . . . | . . . | . . . | . . . | . . . | . . . | . . . | . . . |
| 24 | . . . | . . . | . . . | . . . | . . . | . . . | . . . | . . . | . . . | . . . | . . . | . . . | . . . |
| 25 | 887 | 911 | 916 | 930 | 925 | 920 | 920 | 920 | 916 | 920 | 916 | 916 | 916 |
| 26 | 791 | 834 | 868 | 901 | 896 | 882 | 887 | 882 | 887 | 892 | 896 | 872 | 871 |
| 27 | 887 | 882 | 887 | 896 | 892 | 896 | 896 | 896 | 892 | 896 | 892 | 896 | 887 |
| 28 | 898 | 921 | 914 | 917 | 916 | 924 | 917 | 930 | 924 | 908 | 882 | 865 | 899 |
| 29 | 640 | 639 | 589 | 616 | 629 | 638 | 645 | 643 | 637 | 631 | 624 | 627 | 738 |
| 30 | 612 | 645 | 648 | 636 | 650 | 649 | 656 | 664 | 653 | 661 | 644 | 629 | 626 |
| 31 | 710 | 704 | 678 | 686 | 675 | 679 | 677 | 689 | 688 | 678 | 704 | 702 | 660 |
| Mean . | 775 | 791 | 786 | 797 | 798 | 798 | 800 | 803 | 800 | 798 | 794 | 787 | 789 |

TABLE VII—Continued—*Ver*

The figures given in the table are millionths of a dyne, which,

.581000 +

| Day. | A. M. | | | | | | | | | | | |
|---|---|---|---|---|---|---|---|---|---|---|---|---|
| | 1. | 2. | 3. | 4. | 5. | 6. | 7. | 8. | 9. | 10. | 11. | 12. |
| 1 | 699 | 698 | 696 | 715 | 703 | 712 | 695 | 664 | 611 | 590 | 540 | 572 |
| 2 | 634 | 642 | 650 | 648 | 647 | 646 | 644 | 637 | 612 | 592 | 575 | 549 |
| 3 | 678 | 639 | 656 | 664 | 658 | 719 | 679 | 667 | 657 | 632 | 582 | 575 |
| 4 | 593 | 602 | 585 | 588 | 587 | 591 | 599 | 616 | 596 | 571 | 559 | 562 |
| 5 | 657 | 666 | 659 | 676 | 675 | 684 | 682 | 699 | 678 | 653 | 641 | 629 |
| 6 | 620 | 648 | 675 | 692 | 686 | 685 | 669 | 671 | 603 | 583 | 557 | 555 |
| 7 | 977 | 995 | 969 | 986 | 985 | 1013 | 1006 | 990 | 965 | 978 | 928 | . . . |
| 8 | . . . | . . . | . . . | . . . | . . . | . . . | . . . | . . . | . . . | . . . | . . . | . . . |
| 9 | 499 | 484 | 496 | 509 | 532 | 555 | 558 | 570 | 564 | 534 | 494 | 487 |
| 10 | 366 | 365 | 363 | 347 | 346 | 355 | 348 | 322 | 297 | 277 | 255 | 258 |
| 11 | 272 | 280 | 283 | 310 | 304 | 313 | 316 | 314 | 293 | 268 | 257 | 260 |
| 12 | 307 | 315 | 313 | 326 | 334 | 343 | 346 | 353 | 319 | 289 | 282 | 299 |
| 13 | . . . | . . . | . . . | | | | . . . | . . . | . . . | . . . | . . . | . . . |
| 14 | . . . | . . . | . . . | . . . | . . . | . . . | . . . | . . . | . . . | . . . | . . . | . . . |
| 15 | 772 | 787 | 774 | 807 | 842 | 867 | 887 | 868 | 797 | 797 | 750 | 746 |
| 16 | . . . | . . . | | | | | | | | | | |
| 17 | 078 | 070 | 056 | 023 | 038 | 025 | 040 | (—)037 | (--)065 | (—)035 | (--)091 | (--)115 |
| 18 | 067 | 083 | 098 | 103 | 075 | 091 | 120 | 096 | 078 | 060 | 047 | 047 |
| 19 | 061 | 062 | 054 | 063 | 074 | 087 | 095 | 071 | 058 | 049 | 041 | 017 |
| 20 | 171 | 128 | 149 | 110 | 116 | 160 | 176 | 176 | 157 | 139 | 087 | 068 |
| 21 | 141 | 142 | 147 | 152 | 153 | 169 | 179 | 184 | 161 | 133 | 115 | 081 |
| 22 | . . . | . . . | . . . | . . . | . . . | . . . | . . . | . . . | . . . | . . . | . . . | . . . |
| 23 | . . . | . . . | . . . | . . . | . . . | . . . | . . . | . . . | . . . | . . . | | . . . |
| 24 | . . . | . . . | . . . | . . . | . . . | . . . | . . . | | . . . | . . . | . . . | . . . |
| 25 | 270 | 257 | 258 | 234 | 225 | 221 | 203 | 184 | 180 | 152 | 062 | (—)010 |
| 26 | 159 | 169 | 175 | 180 | 181 | 182 | 173 | 149 | 131 | 137 | 109 | 047 |
| 27 | 009 | 010 | 011 | 030 | 036 | 037 | 042 | 033 | 034 | (—)008 | (--)022 | (—)041 |
| 28 | 089 | 090 | 091 | 086 | 087 | 088 | 075 | 046 | 042 | 043 | 030 | 025 |
| Mean . | 387 | 387 | 388 | 393 | 394 | 407 | 406 | 394 | 370 | 354 | 324 | 281 |

*tical force, February,* 1889.

added to .581000 dyne, give the vertical force in C. G. S. units.

.581000 +

| Day. | P. M | | | | | | | | | | | | Mean. |
|---|---|---|---|---|---|---|---|---|---|---|---|---|---|
| | 1. | 2. | 3. | 4. | 5. | 6. | 7. | 8. | 9. | 10. | 11. | 12. | |
| 1 | 571 | 642 | 635 | 614 | 608 | 607 | 610 | 637 | 631 | 634 | 632 | 639 | 640 |
| 2 | 668 | 667 | 680 | 668 | 658 | 666 | 669 | 701 | 685 | 698 | 681 | 684 | 650 |
| 3 | 535 | 587 | 600 | 626 | 640 | 624 | 627 | 635 | 634 | 633 | 616 | 610 | 632 |
| 4 | 580 | 589 | 601 | 618 | 608 | 592 | 600 | 617 | 630 | 633 | 636 | 648 | 600 |
| 5 | 672 | 666 | 654 | 633 | 618 | 597 | 586 | 589 | 588 | 601 | 599 | 611 | 646 |
| 6 | 847 | 851 | 873 | 866 | 865 | 864 | 900 | 932 | 988 | 1007 | 1005 | 983 | 776 |
| 7 | ... | ... | ... | ... | ... | ... | ... | ... | ... | ... | ... | ... | 983 |
| 8 | 619 | 652 | 640 | 600 | 556 | 550 | 548 | 560 | 550 | 544 | 547 | 525 | 574 |
| 9 | 477 | 476 | 488 | 452 | 427 | 398 | 381 | 384 | 383 | 381 | 374 | 372 | 470 |
| 10 | 262 | 309 | 360 | 334 | 299 | 260 | 263 | 275 | 260 | 272 | 270 | 268 | 305 |
| 11 | 273 | 306 | 332 | 330 | 320 | 314 | 307 | 305 | 299 | 302 | 300 | 308 | 299 |
| 12 | ... | ... | ... | ... | ... | ... | ... | ... | ... | ... | ... | ... | 309 |
| 13 | ... | ... | ... | ... | ... | ... | ... | ... | ... | ... | ... | ... | ... |
| 14 | 733 | 737 | 720 | 733 | 722 | 720 | 766 | 785 | 784 | 767 | 756 | 774 | 750 |
| 15 | ... | ... | ... | ... | ... | ... | ... | ... | ... | ... | ... | ... | 808 |
| 16 | 093 | 085 | 090 | 095 | 096 | 102 | 093 | 089 | 090 | 091 | 092 | 092 | 092 |
| 17 | 097 | 131 | 161 | 156 | 181 | 178 | 227 | 241 | 213 | 195 | 124 | 076 | 103 |
| 18 | 052 | 063 | 136 | 165 | 151 | 143 | 149 | 144 | 130 | 103 | 104 | 065 | 099 |
| 19 | 056 | 062 | 077 | 125 | 145 | 156 | 147 | 186 | 206 | 193 | 198 | 179 | 103 |
| 20 | 079 | 094 | 134 | 153 | 173 | 184 | 137 | 137 | 147 | 143 | 144 | 144 | 136 |
| 21 | ... | ... | ... | ... | ... | ... | ... | ... | ... | ... | ... | ... | 146 |
| 22 | ... | ... | ... | ... | ... | ... | ... | ... | ... | ... | ... | ... | ... |
| 23 | ... | ... | ... | ... | ... | ... | ... | ... | ... | ... | ... | ... | ... |
| 24 | 184 | 200 | 249 | 239 | 226 | 227 | 242 | 252 | 262 | 278 | 279 | 264 | 242 |
| 25 | ... | ... | ... | ... | ... | 163 | 131 | 174 | 165 | 161 | 158 | 153 | 181 |
| 26 | 076 | 073 | 098 | 112 | 065 | 047 | 038 | 038 | 054 | 079 | 056 | 075 | 108 |
| 27 | (—)006 | 081 | 125 | 160 | 145 | 118 | 138 | 143 | 129 | 116 | 112 | 083 | 063 |
| 28 | 117 | 123 | 148 | 134 | 135 | 150 | 161 | 161 | 195 | 196 | 149 | 101 | 107 |
| Mean | 349 | 370 | 390 | 391 | 382 | 365 | 363 | 380 | 382 | 382 | 373 | 364 | 374 |

9230——5

TABLE VII—Continued—*Ver*

The figures given in the table are millionths of a dyne, which,

.581000 +

| Day. | A. M. | | | | | | | | | | | |
|---|---|---|---|---|---|---|---|---|---|---|---|---|
| | 1. | 2. | 3. | 4. | 5. | 6. | 7. | 8. | 9. | 10. | 11. | 12. |
| 1 | 163 | 221 | 237 | 242 | 238 | 229 | 197 | 187 | 183 | 170 | 156 | 104 |
| 2 | 243 | 230 | 202 | 236 | 241 | 242 | 248 | 248 | 240 | 212 | 165 | 122 |
| 3 | 223 | 200 | 186 | 191 | 192 | 203 | 204 | 185 | 138 | 115 | 092 | 111 |
| 4 | 270 | 266 | 286 | 291 | 301 | 298 | 299 | 279 | 261 | 257 | 254 | 249 |
| 5 | 259 | 260 | 280 | 280 | 281 | 263 | 269 | 254 | 255 | 261 | 248 | 238 |
| 6 | 339 | 206 | 183 | 279 | 208 | 151 | 200 | 243 | 259 | 284 | 290 | 266 |
| 7 | 156 | 109 | 177 | 321 | 312 | 318 | 353 | 357 | 349 | 350 | 308 | 274 |
| 8 | 399 | 371 | 344 | 363 | 383 | 403 | 428 | 428 | 410 | 377 | 306 | 268 |
| 9 | 355 | 351 | 342 | 352 | 353 | 359 | 360 | 355 | 342 | 309 | 238 | 214 |
| 10 | 358 | 364 | 365 | 370 | 371 | 372 | 387 | 373 | 355 | 303 | 260 | 246 |
| 11 | 352 | 349 | 350 | 354 | 365 | 366 | 372 | 372 | 344 | 326 | 312 | 312 |
| 12 | 303 | 319 | 329 | 334 | 345 | 350 | 342 | 327 | 309 | 267 | 230 | 186 |
| 13 | 230 | 193 | 179 | 155 | 161 | 162 | 182 | 187 | 188 | 179 | 176 | 176 |
| 14 | 229 | 273 | 279 | 274 | 280 | 295 | 301 | 301 | 302 | 293 | 304 | 323 |
| 15 | 271 | 272 | 273 | 273 | 279 | 289 | 305 | 295 | 301 | 292 | 264 | 245 |
| 16 | 246 | 247 | 243 | 267 | 273 | 283 | 299 | 279 | 280 | 267 | 230 | 225 |
| 17 | 244 | 245 | 246 | 246 | 247 | 244 | 259 | 249 | 198 | 122 | 118 | 118 |
| 18 | 306 | 283 | 284 | 284 | 261 | 242 | 267 | 258 | 225 | 159 | 126 | 150 |
| 19 | 199 | 209 | 220 | 225 | 226 | 241 | 247 | 237 | 210 | 187 | 149 | 144 |
| 20 | 246 | 251 | 252 | 252 | 258 | 274 | 289 | 265 | 252 | 248 | 215 | 206 |
| 21 | 273 | 249 | 249 | 254 | 254 | 259 | 259 | 249 | 220 | 196 | 139 | 124 |
| 22 | 259 | 196 | 134 | 187 | 244 | 288 | 321 | 326 | 321 | 249 | 192 | 201 |
| 23 | 278 | 254 | 230 | 230 | 240 | 249 | 254 | 259 | 259 | 206 | 163 | 163 |
| 24 | 168 | 163 | 163 | 158 | 158 | 148 | 134 | 115 | 124 | 100 | 086 | 091 |
| 25 | 206 | 230 | 230 | 230 | 235 | 249 | 225 | 225 | 211 | 144 | 110 | 115 |
| 26 | 115 | 115 | 120 | 120 | 120 | 129 | 139 | 139 | 134 | 048 | (—)005 | 019 |
| 27 | 249 | 196 | 163 | 153 | 168 | 168 | 172 | 129 | 072 | 028 | 028 | 033 |
| 28 | (—)226 | (—)039 | (—)068 | (—)024 | (—)015 | 014 | (—)005 | 014 | 028 | (—)048 | (—)111 | (—)087 |
| 29 | 206 | 177 | 201 | 216 | 225 | 240 | 225 | 206 | 182 | 124 | 100 | 086 |
| 30 | 139 | 134 | 134 | 163 | 177 | 196 | 206 | 201 | 196 | 182 | 192 | 206 |
| 31 | 244 | 240 | 240 | 240 | 235 | 244 | 249 | 249 | 240 | 177 | 144 | 110 |
| Mean | 236 | 230 | 228 | 263 | 246 | 251 | 256 | 251 | 238 | 216 | 177 | 172 |

*tical force, March*, 1889.

added to .581000 dyne, give the vertical force in C. G. S. units.

.581000 +

| Day. | P. M. | | | | | | | | | | | | Mean. |
|---|---|---|---|---|---|---|---|---|---|---|---|---|---|
| | 1. | 2. | 3. | 4. | 5. | 6. | 7. | 8. | 9. | 10. | 11. | 12. | |
| 1 | 191 | 254 | 284 | 294 | 271 | 257 | 263 | 278 | 303 | 284 | 281 | 271 | 232 |
| 2 | 132 | 157 | 153 | 197 | 202 | 242 | 257 | 257 | 253 | 250 | 230 | 236 | 216 |
| 3 | 217 | 228 | 267 | 287 | 288 | 279 | 299 | 304 | 305 | 306 | 312 | 297 | 226 |
| 4 | 235 | 212 | 223 | 228 | 234 | 230 | 235 | 255 | 261 | 276 | 282 | 258 | 260 |
| 5 | 297 | 336 | 313 | 284 | 295 | 305 | 316 | 330 | 384 | 428 | 429 | 381 | 302 |
| 6 | 295 | 344 | 384 | 374 | 370 | 357 | 358 | 377 | 373 | 355 | 342 | 323 | 298 |
| 7 | 294 | 310 | 354 | 359 | 374 | 370 | 371 | 386 | 391 | 407 | 408 | 393 | 325 |
| 8 | 303 | 337 | 362 | 333 | 349 | 350 | 360 | 351 | 357 | 353 | 359 | 354 | 360 |
| 9 | 253 | 259 | 270 | 284 | 300 | 296 | 306 | 330 | 341 | 347 | 348 | 348 | 317 |
| 10 | 295 | 325 | 336 | 302 | 308 | 309 | 310 | 320 | 335 | 336 | 347 | 342 | 333 |
| 11 | 333 | 329 | 315 | 311 | 307 | 308 | 318 | 357 | 387 | 383 | 355 | 321 | 342 |
| 12 | 202 | 227 | 242 | 276 | 301 | 331 | 317 | 288 | 270 | 262 | 258 | 248 | 286 |
| 13 | 201 | 230 | 270 | 327 | 324 | 334 | 330 | 378 | 427 | 337 | 338 | 357 | 251 |
| 14 | 295 | 292 | 259 | 259 | 284 | 290 | 281 | 267 | 258 | 269 | 260 | 246 | 280 |
| 15 | 261 | 262 | 272 | 282 | 283 | 284 | 294 | 304 | 300 | 316 | 297 | 254 | 282 |
| 16 | 221 | 222 | 228 | 252 | 243 | 249 | 250 | 264 | 261 | 257 | 263 | 301 | 256 |
| 17 | 100 | 158 | 222 | 260 | 434 | 757 | 690 | 762 | 610 | 529 | 429 | 314 | 325 |
| 18 | 204 | 196 | 206 | 206 | 212 | 227 | 200 | 195 | 201 | 226 | 179 | 198 | 221 |
| 19 | 126 | 137 | 133 | 181 | 206 | 221 | 237 | 251 | 262 | 253 | 254 | 245 | 209 |
| 20 | 173 | 188 | 233 | 257 | 291 | 292 | 293 | 284 | 280 | 290 | 291 | 287 | 257 |
| 21 | 134 | 153 | 192 | 216 | 249 | 268 | 278 | 292 | 297 | 302 | 307 | 302 | 238 |
| 22 | 259 | 278 | 307 | 331 | 364 | 360 | 355 | 350 | 336 | 326 | 312 | 288 | 283 |
| 23 | 177 | 192 | 206 | 211 | 187 | 192 | 158 | 158 | 153 | 158 | 168 | 168 | 205 |
| 24 | 139 | 172 | 177 | 196 | 211 | 182 | 240 | 201 | 216 | 230 | 244 | 230 | 165 |
| 25 | 153 | 148 | 134 | 134 | 139 | 134 | 115 | 120 | 115 | 115 | 115 | 120 | 165 |
| 26 | 067 | 072 | 096 | 076 | 096 | 110 | 110 | 129 | 124 | 144 | . . . | . . . | 101 |
| 27 | 024 | 033· | 038 | 048 | 057 | 086 | 067 | 081 | 091 | 100 | 000 | (—)125 | 086 |
| 28 | 110 | 168 | 321 | 244 | 244 | 249 | 240 | 283 | 230 | 225 | 220 | 220 | 091 |
| 29 | 129 | 153 | 168 | 177 | 192 | 196 | 182 | 177 | 177 | 177 | 177 | 144 | 177 |
| 30 | 249 | 216 | 225 | 192 | 211 | 249 | 249 | 235 | 254 | 264 | 273 | 268 | 209 |
| 31 | 100 | 086 | 139 | 196 | 216 | 211 | 201 | 187 | 187 | 177 | 172 | 163 | 194 |
| Mean . | 199 | 215 | 236 | 244 | 259 | 275 | 274 | 283 | 282 | 280 | 275 | 258 | 243 |

MAGNETIC OBSERVATIONS AT THE U. S. NAVAL OBSERVATORY.

TABLE VII—Continued—*Ver*

The figures given in the table are millionths of a dyne, which,

.580000 +

| Day. | A. M. | | | | | | | | | | | |
|---|---|---|---|---|---|---|---|---|---|---|---|---|
| | 1. | 2. | 3. | 4. | 5. | 6. | 7. | 8. | 9. | 10. | 11. | 12. |
| 1 | 1168 | 1124 | 1100 | 1115 | 1124 | 1129 | 1139 | 1134 | 1100 | 1048 | 1033 | 1019 |
| 2 | 1163 | 1172 | 1187 | 1187 | 1182 | 1182 | 1187 | 1172 | 1148 | 1115 | 1110 | 1105 |
| 3 | 1192 | 1105 | 1120 | 1062 | 1129 | 1172 | 1177 | 1192 | 1163 | 1144 | 1120 | 1100 |
| 4 | 1187 | 1187 | 1182 | 1187 | 1187 | 1192 | 1206 | 1230 | 1254 | 1201 | 1187 | 1187 |
| 5 | 1350 | 1345 | 1350 | 1364 | 1369 | 1388 | 1388 | 1408 | 1422 | 1393 | . . . | . . . |
| 6 | 1014 | 1014 | 1014 | 1014 | 1019 | 1024 | 1019 | 1024 | 976 | 923 | 889 | 932 |
| 7 | 1052 | 1052 | 1048 | 1043 | 1028 | 1028 | 1004 | 985 | 947 | 860 | 832 | 836 |
| 8 | 918 | 716 | 736 | 798 | 894 | 980 | 1000 | 1028 | 990 | 918 | 913 | 956 |
| 9 | 1052 | 1038 | 1043 | 1033 | 1038 | 1052 | 1057 | 1048 | 1067 | 932 | 937 | 908 |
| 10 | 1019 | 1009 | 1000 | 966 | 942 | 942 | 937 | 942 | 928 | 908 | 812 | 803 |
| 11 | 980 | 980 | 980 | 1000 | 1004 | 1014 | 990 | 985 | 952 | 918 | 904 | 908 |
| 12 | 865 | 841 | 832 | 832 | 836 | 846 | 841 | 846 | 822 | 788 | 784 | 793 |
| 13 | 860 | 870 | 875 | 870 | 865 | 865 | 870 | 856 | 836 | 836 | 856 | 894 |
| 14 | 913 | 918 | 927 | 922 | 930 | 940 | 949 | 968 | 948 | 933 | 937 | 966 |
| 15 | 1003 | 997 | 996 | 987 | 1011 | 1030 | 1029 | 1043 | 1013 | 965 | 936 | 964 |
| 16 | 986 | 985 | 985 | 984 | 994 | 1023 | 1026 | 1026 | 1016 | 973 | 976 | 962 |
| 17 | 1033 | 1023 | 1017 | 1017 | 1012 | 1026 | 1030 | 1006 | 966 | 962 | 927 | 893 |
| 18 | 998 | 1003 | 992 | 997 | 996 | 1001 | 1004 | 1009 | 979 | 912 | 849 | 853 |
| 19 | 928 | 914 | 918 | 922 | 931 | 936 | 940 | 935 | 895 | 857 | 842 | 813 |
| 20 | 840 | 845 | 839 | 825 | 814 | 819 | 789 | 785 | 764 | 750 | 754 | 725 |
| 21 | 762 | 729 | 723 | 737 | 751 | 760 | 745 | 716 | 667 | 571 | 565 | 618 |
| 22 | 698 | 708 | 712 | 707 | 720 | 720 | 715 | 719 | 699 | 651 | 631 | 650 |
| 23 | 782 | 787 | 801 | 801 | 747 | 742 | 765 | 775 | 750 | 716 | 710 | 758 |
| 24 | 867 | 882 | 886 | 886 | 880 | 894 | 893 | 869 | 854 | 844 | 795 | 790 |
| 25 | 804 | 804 | 803 | 803 | 792 | 792 | 796 | 753 | 732 | 728 | 731 | 736 |
| 26 | 600 | 676 | 719 | 747 | 770 | 775 | 745 | 745 | 725 | 706 | 662 | 652 |
| 27 | 761 | 757 | 751 | 736 | 764 | 788 | 768 | 758 | 724 | 671 | 665 | 689 |
| 28 | 698 | 741 | 764 | 764 | 782 | 777 | 747 | 704 | 679 | 660 | 645 | 621 |
| 29 | 671 | 681 | 694 | 694 | 703 | 717 | 735 | 721 | 696 | 677 | 676 | 661 |
| 30 | 607 | 603 | 606 | 602 | 610 | 620 | 624 | 614 | 599 | 579 | 545 | 564 |
| Mean . | 896 | 917 | 920 | 920 | 927 | 939 | 937 | 933 | 911 | 905 | 828 | 840 |

*tical force,* **April,** 1889.

added to .580000 dyne, give the vertical force in C. G. S. units.

.580000+

| Day. | P. M. | | | | | | | | | | | | Mean. |
|---|---|---|---|---|---|---|---|---|---|---|---|---|---|
| | 1. | 2. | 3. | 4. | 5. | 6. | 7. | 8. | 9. | 10. | 11. | 12. | |
| 1 | 1028 | 1033 | 1096 | 1134 | 1153 | 1187 | 1177 | 1192 | 1187 | 1187 | 1187 | 1158 | 1123 |
| 2 | 1100 | 1129 | 1158 | 1187 | 1177 | 1182 | 1177 | 1182 | 1192 | 1206 | 1196 | 1201 | 1167 |
| 3 | 1139 | 1192 | 1216 | 1225 | 1249 | 1216 | 1206 | 1196 | 1196 | 1187 | 1192 | 1192 | 1170 |
| 4 | 1244 | 1249 | 1244 | 1278 | 1288 | 1326 | 1312 | 1331 | 1350 | 1364 | 1364 | 1364 | 1254 |
| 5 | 932 | 904 | 904 | 947 | 985 | 1009 | 1004 | 1004 | 1014 | 1019 | 1024 | 1019 | 1161 |
| 6 | 928 | 928 | 942 | 980 | 1014 | 1014 | 1019 | 1048 | 1052 | 1057 | 1062 | 1057 | 998 |
| 7 | 980 | 961 | 1000 | 1014 | 1009 | 1024 | 1014 | 1024 | 1096 | 590 | 1091 | 1072 | 1000 |
| 8 | 985 | 990 | 1004 | 1086 | 1096 | 1105 | 1096 | 1081 | 1076 | 1081 | 1086 | 1076 | 984 |
| 9 | 942 | 980 | 990 | 1009 | 1004 | 1033 | 1057 | 1014 | 1033 | 1033 | 1019 | 1033 | 1015 |
| 10 | 846 | 884 | 894 | 942 | 947 | 956 | 942 | 947 | 961 | 966 | 966 | 976 | 935 |
| 11 | 875 | 860 | 884 | 913 | 923 | 928 | 913 | 894 | 875 | 865 | 875 | 870 | 929 |
| 12 | 827 | 827 | 856 | 860 | 870 | 875 | 894 | 899 | 904 | 899 | 908 | 856 | 850 |
| 13 | 923 | 928 | 932 | 937 | 947 | 942 | 932 | 918 | 918 | 918 | 923 | 923 | 896 |
| 14 | 998 | 984 | 993 | 1006 | 1020 | 1019 | 1014 | 1018 | 1018 | 1004 | 1013 | 1008 | 973 |
| 15 | 930 | 948 | 962 | 1006 | 1005 | 989 | 970 | 974 | 979 | 982 | 982 | 982 | 987 |
| 16 | 1028 | 1028 | 1066 | 1075 | 1084 | 1055 | 1049 | 1044 | 1043 | 1043 | 1033 | 1038 | 1022 |
| 17 | 931 | 936 | 968 | 997 | 1015 | 1020 | 1014 | 1029 | 1037 | 1032 | 1027 | 993 | 996 |
| 18 | 838 | 843 | 890 | 914 | .942 | 942 | 936 | 941 | 945 | 940 | 929 | 934 | 941 |
| 19 | 773 | 769 | 768 | 787 | 824 | 853 | 842 | 842 | 841 | 846 | 831 | 840 | 860 |
| 20 | 729 | 734 | 728 | 781 | 789 | 808 | 788 | 788 | 763 | 787 | 786 | 786 | 784 |
| 21 | 703 | 708 | 722 | 730 | 749 | 735 | 705 | 686 | 704 | 632 | 684 | 703 | 701 |
| 22 | 702 | 750 | 763 | 797 | 791 | 781 | 775 | 780 | 765 | 780 | 783 | 783 | 733 |
| 23 | 892 | 887 | 919 | 910 | 928 | 971 | 951 | 913 | 902 | 840 | 844 | 863 | 831 |
| 24 | 818 | 823 | 851 | 860 | 840 | 821 | 810 | 801 | 805 | 805 | 808 | 808 | 841 |
| 25 | 735 | 807 | 801 | 869 | 925 | 911 | 881 | 847 | 842 | 779 | 788 | 625 | 796 |
| 26 | 694 | 733 | 775 | 809 | 817 | 808 | 792 | 792 | 777 | 772 | 761 | 761 | 742 |
| 27 | 736 | 779 | ... | ... | ... | 753 | 709 | 743 | 756 | 699 | 611 | 602 | 725 |
| 28 | 653 | 711 | 743 | 772 | 757 | 738 | 727 | 727 | 716 | 712 | 677 | 658 | 716 |
| 29 | 579 | 574 | 597 | 611 | 620 | 625 | 609 | 609 | 613 | 618 | 607 | 607 | 650 |
| 30 | 529 | 539 | 591 | 620 | 614 | 619 | 613 | 589 | 578 | 583 | 587 | 582 | 592 |
| Mean . | 867 | 904 | 905 | 934 | 944 | 942 | 931 | 928 | 931 | 921 | 921 | 912 | 913 |

TABLE VII—Continued—*Ver*

The figures given in the table are millionths of a dyne, which,

.580000 +

| Day. | A. M. | | | | | | | | | | | |
|------|-----|-----|-----|-----|-----|-----|-----|-----|-----|-----|-----|-----|
| | 1. | 2. | 3. | 4. | 5. | 6. | 7. | 8. | 9. | 10. | 11. | 12. |
| 1 | 576 | 571 | 566 | 570 | 589 | 593 | 583 | 559 | 543 | 534 | 504 | 518 |
| 2 | 599 | 608 | 607 | 622 | 626 | 640 | 649 | 653 | 643 | 595 | 575 | 584 |
| 3 | 612 | 612 | 611 | 611 | 619 | 629 | 633 | 657 | 661 | 593 | 568 | 544 |
| 4 | 590 | 528 | 536 | 560 | 579 | 583 | 597 | 602 | 605 | 586 | 585 | 571 |
| 5 | 469 | 507 | 542 | 591 | 616 | 641 | 593 | 578 | 565 | 523 | 547 | 524 |
| 6 | 643 | 662 | 687 | 668 | 688 | 675 | 642 | 604 | 567 | 523 | 520 | 559 |
| 7 | 452 | 529 | 573 | 630 | 670 | 680 | 681 | 638 | 634 | 610 | 631 | 689 |
| 8 | 626 | 645 | 646 | 637 | 633 | 644 | 635 | 611 | 607 | 574 | 599 | 604 |
| 9 | 642 | 647 | 662 | 638 | 625 | 621 | 627 | 622 | 594 | 556 | 543 | 534 |
| 10 | 513 | 504 | 505 | 509 | 515 | 531 | 532 | 536 | 513 | 441 | 423 | 410 |
| 11 | 496 | 486 | 492 | 501 | 526 | 542 | 548 | 552 | 553 | 477 | 430 | . . . |
| 12 | 488 | 492 | 479 | 489 | 494 | 510 | 516 | 501 | 454 | 363 | 345 | 379 |
| 13 | 331 | 345 | 351 | 365 | 410 | 478 | 455 | 464 | 422 | 369 | 284 | 295 |
| 14 | 423 | 403 | 409 | 414 | 439 | 450 | 484 | 489 | 447 | 351 | 333 | 353 |
| 15 | 472 | 472 | 468 | 478 | 489 | 499 | 519 | 515 | 492 | 477 | 353 | 311 |
| 16 | 483 | 498 | 508 | 513 | 543 | 573 | 588 | 593 | 546 | 388 | 259 | 294 |
| 17 | 475 | 446 | 443 | 438 | 453 | 473 | 503 | 522 | 495 | 346 | 232 | 252 |
| 18 | 376 | 429 | 430 | 382 | 388 | 456 | 452 | 447 | 395 | 338 | 286 | 364 |
| 19 | 406 | 368 | 297 | 321 | 346 | 352 | 358 | 367 | 359 | 363 | 331 | 346 |
| 20 | 509 | 494 | 510 | 510 | 520 | 526 | 561 | 570 | 523 | 466 | 404 | 420 |
| 21 | 533 | 529 | 515 | 506 | 526 | 532 | 547 | 533 | 558 | 462 | 448 | 478 |
| 22 | 295 | 319 | . . . | . . . | . . . | . . . | . . . | . . . | . . . | 506 | 507 | 508 |
| 23 | 633 | 633 | 634 | 634 | 659 | 679 | 665 | 656 | 638 | 590 | 591 | 620 |
| 24 | 705 | 700 | 701 | 706 | 707 | 694 | 685 | 666 | 619 | 571 | 596 | 616 |
| 25 | 553 | 553 | 543 | 537 | 546 | 541 | 549 | 544 | 539 | 505 | 485 | 503 |
| 26 | 447 | 375 | 398 | 422 | 296 | 305 | 342 | 275 | 284 | 260 | 273 | 307 |
| 27 | 524 | 462 | 451 | 533 | 561 | 574 | 559 | 563 | 519 | 409 | 393 | 450 |
| 28 | 548 | 548 | 527 | 527 | 555 | 554 | 582 | 563 | 567 | 552 | 556 | 508 |
| 29 | 452 | 457 | 465 | 489 | 508 | 507 | 506 | 506 | 529 | 481 | 470 | 459 |
| 30 | 451 | 442 | 446 | 455 | 473 | 468 | 452 | 433 | 418 | 432 | 345 | 297 |
| 31 | 394 | 375 | 350 | 360 | 368 | 372 | 385 | 357 | 293 | 269 | 268 | 262 |
| Mean . | 507 | 504 | 512 | 520 | 532 | 544 | 548 | 539 | 519 | 468 | 441 | 452 |

*tical force,* May, 1889.

added to .580000 dyne, give the vertical force in C. G. S. units.

.580000 +

| Day. | P. M. | | | | | | | | | | | | Mean. |
|---|---|---|---|---|---|---|---|---|---|---|---|---|---|
| | 1. | 2. | 3. | 4. | 5. | 6. | 7. | 8. | 9. | 10. | 11. | 12 | |
| 1 | 594 | 633 | 665 | . . . | . . . | . . . | 577 | 582 | 595 | 595 | 590 | 589 | 577 |
| 2 | 650 | 684 | 688 | 693 | 677 | 658 | 638 | 633 | 623 | 622 | 622 | 613 | 633 |
| 3 | 567 | 582 | 595 | 638 | 637 | 628 | 598 | 603 | 626 | 631 | 630 | 610 | 612 |
| 4 | 603 | 671 | 679 | 694 | 688 | 673 | 653 | 639 | 639 | 643 | 633 | 561 | 612 |
| 5 | 630 | 693 | 780 | 819 | 824 | 777 | 750 | 726 | 717 | 707 | 689 | 676 | 677 |
| 6 | 560 | 613 | 647 | 686 | 682 | 669 | 674 | 670 | 675 | 666 | 532 | 403 | 621 |
| 7 | 733 | 729 | 754 | 763 | 759 | 722 | 694 | 675 | 671 | 676 | 677 | 640 | 663 |
| 8 | 615 | 639 | 664 | 702 | 689 | 661 | 638 | 614 | 610 | 610 | 611 | 612 | 630 |
| 9 | 506 | 521 | 560 | 589 | 599 | 586 | 558 | 558 | 554 | 545 | 541 | 531 | 582 |
| 10 | 416 | 464 | 479 | 517 | 523 | 529 | 506 | 492 | 488 | 488 | 498 | 499 | 497 |
| 11 | 475 | 460 | 471 | 485 | 501 | 507 | 503 | 493 | 489 | 499 | 495 | 491 | 499 |
| 12 | 342 | 356 | 471 | 415 | 416 | 422 | 423 | 418 | 424 | 424 | 324 | 277 | 426 |
| 13 | 329 | 387 | 455 | 450 | 451 | 447 | 444 | 434 | 430 | 421 | 422 | 422 | 403 |
| 14 | 402 | 450 | 513 | 508 | 524 | 530 | 502 | 497 | 488 | 479 | 485 | 476 | 452 |
| 15 | 317 | 403 | 462 | 476 | 482 | 488 | 465 | 465 | 466 | 476 | 477 | 487 | 459 |
| 16 | . . . | . . . | . . . | . . . | . . . | 590 | 562 | 543 | 535 | 515 | 492 | 474 | 500 |
| 17 | 272 | 363 | 407 | 431 | 432 | 438 | 396 | 391 | 392 | 397 | 408 | 375 | 407 |
| 18 | 350 | 365 | 399 | 447 | 458 | 430 | 412 | 407 | 399 | 408 | 409 | 410 | 401 |
| 19 | 357 | 390 | 468 | 516 | 527 | 571 | 558 | 534 | 539 | 530 | 516 | 508 | 426 |
| 20 | 450 | 517 | 566 | 575 | 572 | 563 | 526 | 526 | 527 | 531 | 528 | 533 | 548 |
| 21 | 465 | 522 | 566 | 576 | 591 | 578 | 565 | 569 | 590 | 604 | 595 | 414 | 533 |
| 22 | 610 | 644 | 683 | . . . | . . . | . . . | 677 | 653 | 649 | 639 | 645 | 636 | 569 |
| 23 | 650 | 698 | 776 | 795 | 782 | 725 | 707 | 702 | 722 | 689 | 695 | 699 | 678 |
| 24 | 598 | 636 | 609 | 618 | 643 | 644 | 592 | 578 | 560 | 569 | 561 | 562 | 631 |
| 25 | 425 | 468 | 491 | 525 | 505 | 470 | 450 | 445 | 435 | 439 | 438 | 443 | 497 |
| 26 | 450 | 502 | 612 | 645 | 673 | 601 | 600 | 600. | 590 | 570 | 569 | 525 | 455 |
| 27 | 497 | 502 | 525 | 544 | 586 | 580 | 570 | 565 | 564 | 545 | 553 | 539 | 524 |
| 28 | 425 | 411 | 472 | 530 | 524 | 524 | 499 | 480 | 465 | 465 | 464 | 458 | 513 |
| 29 | 463 | 473 | 501 | 520 | 533 | 523 | 498 | 478 | 477 | 477 | 467 | 452 | 487 |
| 30 | 233 | 296 | 371 | 415 | 452 | 514 | 480 | 446 | 440 | 373 | 382 | 400 | 413 |
| 31 | 165 | 209 | 241 | 280 | 303 | 306 | 301 | 315 | 314 | 304 | 289 | 274 | 306 |
| Mean . | 472 | 509 | 552 | 566 | 573 | 564 | 546 | 540 | 538 | 533 | 524 | 503 | 521 |

TABLE VII—Continued—*Ver*

The figures given in the table are millionths of a dyne, which,

.579000 +

| Day. | A. M. | | | | | | | | | | | |
|------|-----|-----|-----|-----|-----|-----|-----|-----|-----|-----|-----|-----|
|      | 1. | 2. | 3. | 4. | 5. | 6. | 7. | 8. | 9. | 10. | 11. | 12. |
| 1 | 1273 | 1259 | 1234 | 1228 | 1217 | 1255 | 1269 | 1297 | 1302 | 1272 | 1253 | . . . |
| 2 | 1379 | 1374 | 1378 | 1377 | 1381 | 1375 | 1375 | 1379 | 1331 | 1272 | 1220 | 1199 |
| 3 | 1331 | 1283 | 1325 | 1329 | 1356 | 1355 | 1365 | 1374 | 1345 | 1320 | 1320 | 1305 |
| 4 | 1437 | 1456 | 1464 | 1459 | 1467 | 1466 | 1461 | 1470 | 1470 | 1426 | 1373 | 1329 |
| 5 | 1359 | 1359 | 1363 | 1362 | 1366 | 1365 | 1355 | 1349 | 1345 | 1310 | 1276 | 1248 |
| 6 | 1355 | 1379 | 1368 | 1377 | 1385 | 1399 | 1404 | 1431 | 1422 | 1358 | 1339 | 1372 |
| 7 | 1383 | 1388 | 1387 | 1391 | 1399 | 1398 | 1393 | 1388 | 1354 | 1310 | 1228 | 1233 |
| 8 | 1311 | 1316 | 1315 | 1314 | 1323 | 1336 | 1331 | 1311 | 1321 | 1253 | 1166 | 1208 |
| 9 | 1268 | 1268 | 1262 | 1237 | 1169 | 1235 | 1273 | 1263 | 1229 | 1118 | 1108 | 1151 |
| 10 | 1321 | 1326 | 1325 | 1333 | 1342 | 1346 | 1341 | 1321 | 1282 | 1185 | 1152 | 1175 |
| 11 | 1273 | 1273 | 1262 | 1289 | 1322 | 1331 | 1321 | 1291 | 1282 | 1228 | 1213 | 1213 |
| 12 | 1273 | 1277 | 1286 | 1286 | 1313 | 1307 | 1312 | 1301 | 1282 | 1209 | 1199 | 1242 |
| 13 | 1358 | 1354 | 1348 | 1356 | 1379 | 1388 | 1359 | 1349 | 1349 | 1309 | 1256 | 1199 |
| 14 | 1186 | 1085 | 1031 | 1011 | 967 | 880 | 889 | 1042 | 1047 | 1036 | 1055 | 1150 |
| 15 | 1329 | 1284 | 1287 | 1295 | 1312 | 1318 | 1331 | 1324 | 1298 | 1266 | 1197 | 1224 |
| 16 | . . . | . . . | . . . | . . . | . . . | . . . | . . | . . . | . . . | . . . | . . . | . . . |
| 17 | 1205 | 1198 | 1201 | 1218 | 1236 | 1257 | 1245 | 1233 | 1183 | 1185 | 1222 | 1234 |
| 18 | 1152 | 1169 | 1167 | 1170 | 1178 | 1184 | 1168 | 1156 | 1140 | 1089 | 1049 | 1027 |
| 19 | 1171 | 1164 | 1182 | 1194 | 1235 | 1232 | 1230 | 1257 | 1265 | 1223 | 1207 | 1190 |
| 20 | 1220 | 1223 | 1211 | 1195 | 1207 | 1204 | 1202 | 1191 | 1184 | 1195 | 1150 | 1138 |
| 21 | 1061 | 1059 | 1043 | 916 | 962 | 1031 | 1029 | 1042 | 1044 | 998 | 1006 | 1009 |
| 22 | 917 | 896 | 889 | 921 | 938 | 940 | 928 | 902 | 895 | 888 | 842 | 797 |
| 23 | 907 | 872 | 846 | 877 | 895 | 940 | 962 | 969 | 948 | 959 | 972 | 946 |
| 24 | 971 | 964 | 952 | 950 | 958 | 969 | 967 | 946 | 944 | 907 | 872 | 879 |
| 25 | 874 | 872 | 885 | 902 | 929 | 935 | 914 | 912 | 910 | 878 | 780 | 817 |
| 26 | 903 | 901 | 899 | 907 | 914 | 916 | 914 | 902 | 886 | 859 | 804 | 797 |
| 27 | 769 | 763 | 756 | 759 | 776 | 773 | 800 | 802 | 781 | 749 | 608 | 616 |
| 28 | 697 | 705 | 693 | 706 | 704 | 715 | 713 | 711 | 675 | 658 | . . . | . . . |
| 29 | 764 | 773 | 771 | 783 | 786 | 788 | 795 | 774 | 758 | 745 | 719 | 688 |
| 30 | 674 | 677 | 675 | 668 | 680 | 696 | 699 | 692 | 671 | 639 | 604 | 572 |
| Mean . | 1142 | 1137 | 1131 | 1131 | 1141 | 1149 | 1150 | 1151 | 1136 | 1098 | 1078 | 1073 |

*tical force, June,* 1889.

added to .579000 dyne, give the vertical force in C. G. S. units.

.579000 +

| Day. | P. M. | | | | | | | | | | | | Mean. |
|---|---|---|---|---|---|---|---|---|---|---|---|---|---|
| | 1. | 2. | 3. | 4. | 5. | 6. | 7. | 8. | 9. | 10. | 11. | 12. | |
| 1 | 1285 | 1361 | 1395 | 1427 | 1436 | 1431 | 1425 | 1421 | 1424 | 1396 | 1380 | 1409 | 1333 |
| 2 | 1219 | 127c | 1327 | 1365 | 1379 | 1373 | 1358 | 1367 | 1376 | 1376 | 1370 | 1346 | 1340 |
| 3 | 1472 | 1520 | 1548 | 1587 | 1614 | 1590 | 1537 | 1508 | 1497 | 1492 | 1439 | 1404 | 1426 |
| 4 | 1275 | 1285 | 1317 | 1380 | 1393 | 1397 | 1386 | 1372 | 1366 | 1361 | 1355 | 1360 | 1397 |
| 5 | 1285 | 1343 | 1361 | 1361 | 1379 | 1360 | 1340 | 1344 | 1363 | 1372 | 1366 | 1361 | 1346 |
| 6 | 1342 | 1356 | 1413 | 1437 | 1417 | 1401 | 1400 | 1376 | 1385 | 1385 | 1374 | 1384 | 1386 |
| 7 | 1261 | 1294 | 1341 | 1365 | 1364 | 1360 | 1330 | 1325 | 1319 | 1319 | 1318 | 1317 | 1340 |
| 8 | 1251 | 1299 | 1317 | 1336 | 1321 | 1305 | 1290 | 1271 | 1274 | 1279 | 1288 | 1273 | 1292 |
| 9 | 1213 | 1294 | 1312 | 1360 | 1374 | 1355 | 1349 | 1334 | 1338 | 1338 | 1337 | 1322 | 1271 |
| 10 | 1193 | 1241 | 1288 | 1312 | 1325 | 1315 | 1299 | 1294 | 1298 | 1284 | 1273 | 1273 | 1285 |
| 11 | 1280 | 1184 | 1307 | 1355 | 1383 | 1364 | 1325 | 1310 | 1295 | 1295 | 1289 | 1288 | 1291 |
| 12 | 1289 | 1289 | . . . | 1321 | 1335 | 1343 | 1256 | 1337 | 1351 | 1365 | 1364 | 1355 | 1300 |
| 13 | . . . | . . . | . . . | . . . | . . . | . . . | . . . | . . . | 1251 | 1271 | 1207 | 1201 | 1308 |
| 14 | 1303 | 1365 | 1393 | 1393 | 1411 | 1482 | 1486 | 1481 | 1461 | 1437 | 1412 | 1388 | 1225 |
| 15 | 1106 | 1123 | 1121 | 1392 | 1154 | 1162 | 1184 | 1172 | 1175 | . . . | . . . | . . . | 1241 |
| 16 | 1144 | 1180 | 1221 | 1243 | 1250 | 1238 | 1217 | 1206 | 1180 | 1202 | 1228 | 1220 | 1211 |
| 17 | 1154 | 1138 | 1126 | 1172 | 1188 | 1182 | 1151 | 1134 | 1128 | 1129 | 1142 | 1149 | 1184 |
| 18 | 1005 | 1003 | 1049 | 1134 | 1159 | 1143 | 1136 | 1149 | 1147 | 1139 | 1156 | 1173 | 1127 |
| 19 | 1207 | 1229 | 1227 | 1258 | 1271 | 1264 | 1267 | 1250 | 1195 | 1226 | 1224 | 1222 | 1225 |
| 20 | 1068 | 1105 | 1122 | 1129 | 1099 | 1077 | 1085 | 1073 | 1067 | 1064 | 1066 | 1060 | 1139 |
| 21 | . . . | 984 | 1002 | 1057 | 1094 | 1058 | 1022 | 1035 | 922 | 943 | 970 | 973 | 1011 |
| 22 | 813 | 864 | 785 | 908 | 906 | 914 | 912 | 895 | 898 | 895 | 903 | 910 | 990 |
| 23 | 1053 | 1061 | 1088 | 1076 | 1064 | 1024 | 993 | 982 | 960 | 962 | 965 | 973 | 971 |
| 24 | 833 | 855 | 872 | 909 | 926 | 929 | 922 | 915 | 908 | 891 | 889 | 906 | 918 |
| 25 | 905 | 922 | 939 | 966 | 974 | 953 | 984 | 933 | 936 | 934 | 927 | 895 | 912 |
| 26 | 761 | 768 | 824 | 851 | 868 | 847 | 816 | 794 | 787 | 785 | 778 | 776 | 844 |
| 27 | 656 | 654 | 686 | 736 | 744 | 747 | 740 | 723 | 711 | 714 | 707 | 699 | 728 |
| 28 | 699 | 678 | 762 | 871 | 854 | 900 | 908 | 866 | 850 | 843 | 827 | 795 | 765 |
| 29 | 652 | 664 | 710 | 723 | 711 | 699 | 712 | 709 | 707 | 700 | 683 | 677 | 733 |
| 30 | 613 | 688 | 715 | 732 | 759 | 738 | 716 | 694 | 692 | 700 | 703 | 695 | 684 |
| Mean. | 1083 | 1104 | 1127 | 1178 | 1178 | 1171 | 1157 | 1147 | 1142 | 1141 | 1136 | 1131 | 1134 |

TABLE VII—Continued—*Ver*

The figures given in the table are millionths of a dyne, which,

.579000 +

| Day. | A. M. | | | | | | | | | | | |
|---|---|---|---|---|---|---|---|---|---|---|---|---|
| | 1. | 2. | 3. | 4. | 5. | 6. | 7. | 8. | 9. | 10. | 11. | 12. |
| 1 | 697 | 690 | 612 | 561 | 650 | 657 | 662 | 638 | 617 | 600 | 565 | 547 |
| 2 | . . . | . . . | . . . | . . . | . . . | . . . | . . . | . . . | . . . | . . . | . . . | . . . |
| 3 | . . . | . . . | . . . | . . . | . . . | . . . | . . . | . . . | . . . | . . . | . . . | . . . |
| 4 | 592 | 595 | 593 | 600 | 597 | 600 | 603 | 591 | 508 | 471 | 484 | 520 |
| 5 | 553 | 518 | 521 | 567 | 612 | 624 | 632 | 630 | 580 | 509 | 450 | 501 |
| 6 | 559 | 600 | 617 | 625 | 641 | 639 | 618 | 626 | 604 | 592 | 590 | 569 |
| 7 | 602 | 538 | 560 | 610 | 641 | 649 | 651 | 649 | 619 | 563 | 527 | 525 |
| 8 | . . . | . . . | . . . | . . . | . . . | . . . | . . . | . . . | . . . | . . . | 513 | 520 |
| 9 | 531 | 514 | 512 | 529 | 541 | 544 | 551 | 535 | 514 | 501 | 475 | 454 |
| 10 | 550 | 548 | 546 | 548 | 560 | 572 | 585 | 573 | 499 | 486 | 470 | . . . |
| 11 | 535 | 528 | 531 | 529 | 550 | 548 | 551 | 525 | 480 | 467 | 417 | . . . |
| 12 | 549 | 552 | 545 | 558 | 569 | 567 | 570 | 534 | 499 | 438 | 422 | 429 |
| 13 | 540 | 538 | 541 | 539 | 555 | 549 | 547 | 545 | 523 | 506 | 466 | 468 |
| 14 | 483 | 476 | 479 | 486 | 493 | 510 | 513 | 520 | 528 | 515 | 461 | 478 |
| 15 | 526 | 524 | 522 | 520 | 521 | 534 | 541 | 525 | 523 | 501 | 470 | 444 |
| 16 | 536 | 534 | 542 | 549 | 575 | 621 | 648 | 651 | 629 | 521 | 480 | 450 |
| 17 | 550 | 528 | −133 | −048 | 305 | 384 | 410 | 436 | 496 | 512 | 446 | 482 |
| 18 | 575 | 519 | 517 | 466 | 545 | 538 | 536 | 533 | 515 | 518 | 485 | 468 |
| 19 | 469 | 433 | 440 | 481 | 503 | 524 | 517 | 509 | 506 | 489 | 447 | 444 |
| 20 | 460 | 472 | 503 | 500 | 503 | 500 | 508 | 481 | 482 | 470 | 448 | 422 |
| 21 | 494 | 520 | 504 | 520 | 527 | 524 | 527 | 505 | 492 | 499 | 477 | 474 |
| 22 | 509 | 506 | 514 | 511 | 542 | 549 | 556 | 539 | 531 | 495 | 516 | 546 |
| 23 | 509 | 506 | 495 | 501 | 523 | 520 | 514 | 511 | 508 | 471 | 444 | 455 |
| 24 | 495 | 492 | 524 | 521 | 548 | 550 | 528 | 477 | 470 | 462 | 488 | 470 |
| 25 | 467 | 488 | 496 | 507 | 515 | 521 | 505 | 502 | 528 | 457 | 368 | 384 |
| 26 | 568 | 565 | 558 | 565 | 587 | 594 | 587 | 555 | 552 | 549 | 522 | 466 |
| 27 | 497 | 508 | 516 | 508 | 515 | 532 | 525 | 498 | 476 | 415 | 354 | 356 |
| 28 | 463 | 460 | 463 | 465 | 468 | 470 | 463 | 465 | 466 | 439 | 388 | 371 |
| 29 | 445 | 437 | 406 | 389 | 420 | 427 | 430 | 436 | 428 | 449 | 446 | 434 |
| 30 | 599 | 586 | 555 | 490 | 526 | 562 | 574 | 576 | 582 | 565 | 528 | 463 |
| 31 | 301 | 308 | 407 | 476 | 546 | 567 | 565 | 591 | 506 | 412 | 409 | 463 |
| Mean . | 523 | 517 | 496 | 503 | 539 | 549 | 551 | 541 | 524 | 495 | 467 | 467 |

*tical force, July,* 1889.

added to .579000 dyne, give the vertical force in C. G. S. units.

.579000 +

| Day. | P. M. | | | | | | | | | | | | Mean. |
|------|------|------|------|------|------|------|------|------|------|------|------|------|------|
| | 1. | 2. | 3. | 4. | 5. | 6. | 7. | 8. | 9. | 10. | 11. | 12. | |
| 1 | 560 | 601 | 671 | 679 | 705 | 742 | 724 | 713 | 716 | ... | ... | ... | 648 |
| 2 | 609 | 630 | 647 | 645 | 662 | 650 | 653 | 642 | 649 | 662 | 607 | ... | 641 |
| 3 | 584 | 606 | 585 | 621 | 633 | 636 | 639 | 631 | 619 | 612 | 601 | 594 | 622 |
| 4 | 556 | 550 | 580 | 602 | 624 | 598 | 572 | 560 | 562 | 570 | 573 | 571 | 570 |
| 5 | ... | 588 | 599 | 631 | 653 | 641 | 620 | 612 | 605 | 603 | 611 | ... | 586 |
| 6 | 485 | 497 | 552 | 636 | 678 | 647 | 616 | 618 | 577 | 556 | 573 | 595 | 596 |
| 7 | 528 | 555 | ... | ... | ... | ... | ... | ... | ... | ... | ... | ... | 587 |
| 8 | 509 | 540 | 556 | 578 | 591 | 589 | 568 | 555 | 553 | 541 | 535 | 537 | 549 |
| 9 | 500 | 503 | 538 | 570 | 593 | 580 | 564 | 570 | 573 | 561 | 559 | 553 | 536 |
| 10 | 461 | 498 | 553 | 556 | 573 | 561 | 554 | 552 | 545 | 543 | 535 | 538 | 539 |
| 11 | 447 | 473 | 539 | 585 | 616 | 604 | 597 | 576 | 574 | 566 | 549 | 547 | 528 |
| 12 | 475 | 512 | 558 | 556 | 544 | 531 | 534 | 532 | 540 | 527 | 530 | 533 | 525 |
| 13 | 490 | 512 | 530 | 532 | 540 | 542 | 525 | 523 | 521 | 509 | 502 | 505 | 523 |
| 14 | 461 | 503 | 549 | 551 | 564 | 561 | 535 | 533 | 535 | 542 | 535 | 533 | 514 |
| 15 | 461 | 498 | 496 | 508 | 525 | 537 | 520 | 509 | 520 | 523 | 506 | 509 | 511 |
| 16 | 433 | 489 | 540 | 553 | 564 | 552 | 535 | 533 | 530 | 538 | 546 | 534 | 545 |
| 17 | 561 | 640 | 633 | 669 | 691 | 688 | 647 | 639 | 599 | 615 | 589 | 591 | 493 |
| 18 | 480 | 544 | 595 | 607 | 633 | 650 | 633 | 616 | 590 | 496 | 489 | 457 | 542 |
| 19 | 442 | 487 | 538 | 554 | 576 | 544 | 538 | 544 | 537 | 530 | 499 | 505 | 502 |
| 20 | 409 | 444 | 505 | 516 | 557 | 535 | 548 | 506 | 466 | 492 | 432 | 453 | 484 |
| 21 | 505 | 560 | 572 | 579 | 572 | 574 | 538 | 531 | 524 | 516 | 509 | 506 | 523 |
| 22 | 578 | 584 | 568 | 565 | 563 | 555 | 539 | 531 | 529 | 526 | 519 | 511 | 533 |
| 23 | 458 | 474 | 496 | 498 | 491 | 508 | 506 | 493 | 486 | 493 | 500 | 478 | 493 |
| 24 | 430 | 470 | 502 | 528 | 535 | 542 | 540 | 527 | 501 | 508 | 506 | 503 | 505 |
| 25 | 421 | 456 | 468 | 519 | 583 | 657 | 698 | 677 | 621 | 628 | 583 | 561 | 525 |
| 26 | 435 | 509 | 550 | 548 | 584 | 590 | 584 | 572 | 569 | 523 | 473 | 470 | 545 |
| 27 | 383 | 433 | 455 | 486 | 517 | 519 | 498 | 491 | 483 | 470 | 468 | 461 | 474 |
| 28 | 331 | 337 | 393 | 391 | 426 | 438 | 464 | 458 | 455 | 442 | 430 | 447 | 433 |
| 29 | 518 | 510 | 513 | 574 | 623 | 673 | 681 | 650 | 628 | 625 | 604 | 596 | 514 |
| 30 | 427 | 444 | 523 | 579 | 638 | 645 | 662 | 598 | 580 | 568 | 561 | 524 | 556 |
| 31 | 457 | 458 | 480 | 517 | 571 | 626 | 590 | 646 | 518 | 554 | 552 | 539 | 502 |
| Mean. | 480 | 513 | 543 | 564 | 588 | 591 | 581 | 571 | 557 | 546 | 534 | 524 | 532 |

TABLE VII--Continued--*Ver*

The figures given in the table are millionths of a dyne, which,

.579000 +

| Day. | A. M. | | | | | | | | | | | |
|---|---|---|---|---|---|---|---|---|---|---|---|---|
| | 1. | 2. | 3. | 4. | 5. | 6. | 7. | 8. | 9. | 10. | 11. | 12. |
| 1 | 515 | 494 | 486 | 484 | 514 | 584 | 548 | 555 | 576 | 535 | 375 | 290 |
| 2 | 395 | 451 | 491 | 499 | 520 | 561 | 553 | 541 | 514 | 478 | 500 | 454 |
| 3 | 525 | 538 | 549 | 547 | 558 | 576 | 563 | 561 | 563 | 541 | 467 | 435 |
| 4 | 578 | 572 | 559 | 562 | 592 | 624 | 626 | 595 | 606 | 618 | 573 | 517 |
| 5 | 603 | 606 | 598 | 591 | 607 | 624 | 612 | 605 | 588 | 556 | 554 | 551 |
| 6 | 613 | 625 | 637 | 639 | 656 | 682 | 670 | 682 | 660 | 571 | 554 | 489 |
| 7 | 666 | 693 | 719 | 731 | 766 | 803 | 838 | 827 | 747 | 561 | 425 | 384 |
| 8 | 681 | 674 | 724 | 760 | 815 | 890 | 949 | 952 | 1026 | 1003 | 877 | 730 |
| 9 | 984 | 977 | 998 | 1034 | 1089 | 1173 | 1209 | 1173 | 1103 | 1033 | 920 | 749 |
| 10 | $100_3$ | 1006 | 1003 | 1001 | 1032 | 1058 | 1060 | 1058 | 1002 | 860 | 858 | 932 |
| 11 | ... | ... | ... | ... | ... | ... | ... | ... | ... | ... | ... | ... |
| 12 | 656 | 561 | 707 | 883 | 1059 | 1266 | 1381 | 1407 | 1372 | 1277 | 1213 | 1088 |
| 13 | 339 | 576 | 1022 | 1199 | 1314 | 1340 | 1395 | 1301 | 1206 | 1171 | 1107 | 1012 |
| 14 | 1285 | 1221 | 1157 | 1183 | 1178 | 1204 | 1169 | 1165 | 1160 | 1065 | 880 | 846 |
| 15 | 758 | 754 | 750 | 746 | 741 | 737 | 702 | 608 | 573 | 597 | 533 | ... |
| 16 | 983 | 1099 | 1005 | 1152 | 1357 | 1444 | 1439 | 1435 | 1400 | 1395 | 1361 | 1357 |
| 17 | 1479 | 1475 | 1471 | 1527 | 1552 | 1578 | 1604 | 1539 | 1474 | 1379 | 1285 | 1311 |
| 18 | 1403 | 1429 | 1425 | 1451 | 1507 | 1533 | 1468 | 1343 | 1338 | 1273 | 1359 | 1325 |
| 19 | 1266 | 1353 | 1259 | 1375 | 1401 | 1427 | 1422 | 1327 | 1232 | 1197 | 1163 | 1279 |
| 20 | 1010 | 946 | 882 | 788 | 813 | 839 | 774 | 740 | 705 | 730 | 756 | 722 |
| 21 | 964 | 961 | 987 | 983 | 1008 | 1034 | 999 | 905 | 809 | 684 | 680 | 705 |
| 22 | 707 | 703 | 729 | 755 | 782 | 808 | 803 | 708 | 613 | 548 | 484 | 569 |
| 23 | 631 | 688 | 714 | 740 | 766 | 822 | 787 | 723 | 628 | 592 | 528 | 583 |
| 24 | 916 | 943 | 939 | 965 | 991 | 1017 | 982 | 948 | 883 | 818 | 723 | 688 |
| 25 | 870 | 897 | 893 | 919 | 885 | 911 | 876 | 812 | 747 | 681 | 617 | 702 |
| 26 | 403 | 340 | 215 | 241 | 207 | 113 | 168 | 164 | 159 | 184 | 300 | 386 |
| 27 | 929 | 956 | 982 | 948 | 1094 | 1211 | 1176 | 1172 | 1197 | 1282 | 1248 | 1243 |
| 28 | 1094 | 819 | 1056 | 1112 | 1138 | 1224 | 1220 | 1126 | 1091 | 1056 | 1022 | 1017 |
| 29 | 926 | 954 | 1010 | 1006 | 1002 | 1088 | 1114 | 1050 | 955 | 920 | 886 | 850 |
| 30 | 760 | 757 | 723 | 689 | 715 | 711 | 707 | 733 | 698 | 693 | 719 | 714 |
| 31 | 263 | 230 | 226 | 222 | 308 | 364 | 360 | 386 | 291 | 196 | 073 | (—)084 |
| Mean . | 803 | 810 | 830 | 824 | 899 | 941 | 939 | 905 | 864 | 816 | 768 | 753 |

*tical force,* **August,** 1889.

added to .579000 dyne, give the vertical force in C. G. S. units.

.579000 +

| Day. | P. M. | | | | | | | | | | | | Mean. |
|---|---|---|---|---|---|---|---|---|---|---|---|---|---|
| | 1. | 2. | 3. | 4. | 5. | 6. | 7. | 8. | 9. | 10. | 11. | 12. | |
| 1 | 407 | 481 | 574 | 610 | 655 | 658 | 608 | 595 | 555 | 547 | 531 | 499 | 532 |
| 2 | 417 | 433 | 502 | 524 | 603 | 601 | 589 | 581 | 579 | 576 | 565 | 543 | 520 |
| 3 | 461 | 520 | 575 | 616 | 627 | 606 | 575 | 567 | 565 | 558 | 570 | 567 | 551 |
| 4 | 533 | 583 | 613 | 616 | 637 | 635 | 604 | 611 | 595 | 592 | 594 | 601 | 593 |
| 5 | 605 | 641 | 657 | 631 | 628 | 621 | 610 | 611 | 619 | 616 | 624 | 625 | 608 |
| 6 | 495 | 531 | 600 | 675 | 686 | 655 | 624 | 621 | 619 | 621 | 629 | 650 | 620 |
| 7 | 390 | 464 | 576 | 641 | 667 | 665 | 658 | 660 | 663 | 670 | 696 | 713 | 651 |
| 8 | 731 | 800 | 874 | 983 | 1008 | 992 | 942 | 934 | 951 | 982 | 975 | 977 | 885 |
| 9 | 636 | 729 | 851 | 901 | 956 | 978 | 995 | 1021 | 1029 | 1030 | 1028 | 1006 | 993 |
| 10 | ... | ... | ... | ... | ... | ... | ... | ... | ... | ... | ... | ... | 989 |
| 11 | (—)646 | (—)230 | 007 | 214 | 450 | 536 | 531 | 527 | 673 | 548 | 603 | 690 | 325 |
| 12 | 843 | 838 | 985 | 1193 | 1247 | 1243 | 1208 | 1174 | 1199 | 1285 | 1280 | 1336 | 1113 |
| 13 | 1008 | 1214 | 1511 | 1928 | 1923 | 1829 | 1704 | 1549 | 1484 | 1450 | 1415 | 1351 | 1310 |
| 14 | 782 | 807 | 893 | 919 | 914 | 880 | 845 | 841 | 836 | 832 | 797 | 793 | 981 |
| 15 | 796 | 852 | 968 | 994 | 989 | 985 | 1010 | 1006 | 1031 | 1087 | 1022 | 807 | 828 |
| 16 | 1443 | 1528 | 1554 | 1550 | 1485 | 1451 | 1416 | 1442 | 1407 | 1403 | 1428 | 1454 | 1374 |
| 17 | 1367 | 1512 | 1538 | 1534 | 1469 | 1435 | 1400 | 1366 | 1361 | 1357 | 1412 | 1468 | 1454 |
| 18 | 1351 | 1467 | 1342 | 1428 | 1484 | 1510 | 1264 | 1260 | 1255 | 1221 | 1246 | 1272 | 1373 |
| 19 | 1366 | 1300 | 1176 | 1172 | 1167 | 1163 | 1128 | 1094 | 1119 | 1085 | 1080 | 1046 | 1275 |
| 20 | 718 | 833 | 950 | 1126 | 1242 | 1238 | 1082 | 1018 | 1103 | 1009 | 1004 | 970 | 917 |
| 21 | 670 | 696 | 752 | 809 | 835 | 770 | 705 | 701 | 696 | 722 | 717 | 682 | 811 |
| 22 | 564 | 710 | 797 | 793 | 729 | 725 | 689 | 655 | 650 | 676 | 641 | 666 | 688 |
| 23 | 548 | 604 | 721 | 837 | 893 | 859 | 854 | 911 | 966 | 931 | 896 | 921 | 756 |
| 24 | 773 | 830 | 886 | 912 | 908 | 904 | 839 | 865 | 890 | 885 | 880 | 875 | 886 |
| 25 | 306 | 392 | 479 | 535 | 471 | 376 | 281 | 337 | 362 | 388 | 413 | 438 | 608 |
| 26 | 953 | 1069 | 1155 | 1272 | 1328 | 1204 | 1138 | 1104 | 1099 | 1064 | 1029 | 1084 | 682 |
| 27 | 1328 | 1384 | 1591 | 1707 | 1703 | 1519 | 1394 | 1390 | 1385 | 1289 | 1344 | 1249 | 1280 |
| 28 | 1072 | 1218 | 1305 | 1331 | 1267 | 1323 | 1288 | 1284 | 1156 | 1153 | 938 | 963 | 1136 |
| 29 | 815 | 872 | 928 | 984 | 1010 | 1006 | 911 | 937 | 902 | 806 | 892 | 827 | 944 |
| 30 | 228 | 314 | 310 | 336 | 362 | 358 | 353 | 319 | 314 | 339 | 304 | 269 | 518 |
| 31 | 031 | 268 | 294 | 381 | 407 | 373 | 338 | 334 | 389 | 384 | 319 | 344 | 279 |
| Mean. | 700 | 789 | 862 | 938 | 958 | 936 | 886 | 877 | 882 | 870 | 862 | 856 | 860 |

TABLE VII—Continued—*Ver*

The figures given in the table are millionths of a dyne, which,

.579000 +

| Day. | A. M. | | | | | | | | | | | |
|---|---|---|---|---|---|---|---|---|---|---|---|---|
| | 1. | 2. | 3. | 4. | 5. | 6. | 7. | 8. | 9. | 10. | 11. | 12. |
| 1 | 340 | 335 | 331 | 386 | 382 | 438 | 434 | 429 | 395 | 270 | 146 | 142 |
| 2 | 445 | 440 | 436 | 431 | 427 | 482 | 478 | 473 | 409 | 285 | 220 | 247 |
| 3 | 428 | 484 | 480 | 535 | 531 | 556 | 523 | 488 | 393 | 359 | 355 | 351 |
| 4 | 473 | 469 | 465 | 490 | 486 | 481 | 507 | 442 | 347 | 283 | 339 | 366 |
| 5 | 397 | 423 | 419 | 444 | 410 | 405 | 341 | 276 | 181 | 147 | 203 | 199 |
| 6 | 351 | 377 | 373 | 368 | 364 | 359 | 235 | 170 | 136 | 162 | 158 | 244 |
| 7 | 305 | 301 | 297 | 292 | 318 | 313 | 340 | 274 | 240 | 296 | 383 | 379 |
| 8 | 499 | 435 | 521 | 578 | 574 | 539 | 384 | 409 | 405 | 461 | 518 | 483 |
| 9 | (—)570 | (—)904 | (—)969 | (—)973 | (—)1008 | (—)591 | (—)173 | (—)419 | 028 | 115 | 141 | 167 |
| 10 | 227 | 163 | 309 | 095 | 031 | (—)155 | 142 | 348 | 404 | 400 | 426 | 572 |
| 11 | 693 | 418 | 444 | 591 | 797 | 973 | 909 | 934 | 960 | 866 | 982 | 918 |
| 12 | 978 | 914 | 1031 | 1027 | 932 | 1078 | 1014 | 1069 | 1035 | 1031 | 816 | 782 |
| 13 | . . . | . . . | . . . | . . . | . . . | . . . | . . . | . . . | . . . | . . . | . . . | . . . |
| 14 | . . . | . . . | . . . | . . . | . . . | . . . | . . . | . . . | . . . | . . . | . . . | . . . |
| 15 | 1261 | 1229 | 1196 | 1194 | 1221 | 1340 | 1307 | 1274 | 1302 | 1360 | 1357 | 1204 |
| 16 | 1384 | 1382 | 1349 | 1467 | 1464 | 1402 | 1399 | 1307 | 1364 | 1392 | 1359 | 1176 |
| 17 | 1476 | 1625 | 1562 | 1469 | 1557 | 1495 | 1431 | 1580 | 1547 | 1394 | 1301 | 1269 |
| 18 | . . . | . . . | . . . | . . . | . . . | . . . | . . . | . . . | . . . | . . . | . . . | . . . |
| 19 | 1872 | 1870 | 1867 | 1865 | 1802 | 1860 | 1828 | 1825 | 1792 | 1790 | 1787 | 1755 |
| 20 | 1784 | 1812 | 1839 | 1837 | 1834 | 1892 | 1920 | 1887 | 1884 | 1822 | 1789 | 1787 |
| 21 | 1877 | 1905 | 1902 | 1900 | 1867 | 1865 | 1831 | 1860 | 1766 | 1734 | 1671 | 1669 |
| 22 | 1548 | 1606 | 1663 | 1601 | 1628 | 1626 | 1442 | 1410 | 1437 | 1496 | 1402 | . . . |
| 23 | 1490 | 1427 | 1424 | 1422 | 1419 | 1538 | 1625 | 1653 | 1710 | 1708 | . . . | . . . |
| 24 | 1672 | 1670 | 1637 | 1605 | 1573 | 1480 | 1448 | 1535 | 1593 | 1530 | 1528 | 1525 |
| 25 | 742 | 740 | 767 | 795 | 762 | 760 | 757 | 755 | 752 | 720 | 717 | 715 |
| 26 | 714 | 682 | 648 | 677 | 704 | 702 | 699 | 697 | 664 | 601 | 568 | 566 |
| 27 | 656 | 654 | 651 | 648 | 646 | 673 | 671 | 668 | 666 | 663 | 601 | 599 |
| 28 | 838 | 867 | 864 | 892 | 829 | 887 | 915 | 942 | 910 | 907 | 814 | 842 |
| 29 | 780 | 778 | 775 | 743 | 770 | 799 | 796 | 854 | 760 | 728 | 755 | 753 |
| 30 | 753 | 751 | 748 | 746 | 743 | 741 | 738 | 736 | 733 | 700 | 697 | 635 |
| Mean. | 867 | 846 | 853 | 856 | 854 | 887 | 887 | 884 | 882 | 860 | 809 | 774 |

*tical force, September,* 1889.

added to .579000 dyne, give the vertical force in C. G. S. units.

.579000 +

| Day. | P. M. | | | | | | | | | | | | Mean. |
|---|---|---|---|---|---|---|---|---|---|---|---|---|---|
| | 1. | 2. | 3. | 4. | 5. | 6. | 7. | 8. | 9. | 10. | 11. | 12. | |
| 1 | 317 | 374 | 429 | 455 | 420 | 386 | 351 | 377 | 402 | 398 | 424 | 419 | 366 |
| 2 | 362 | 418 | 413 | 409 | 404 | 400 | 395 | 421 | 416 | 412 | 378 | 403 | 400 |
| 3 | 346 | 463 | 488 | 544 | 479 | 415 | 410 | 406 | 431 | 457 | 453 | 478 | 452 |
| 4 | 331 | 357 | 443 | 318 | 373 | 339 | ·334 | 300 | 325 | 381 | 377 | 372 | 392 |
| 5 | 285 | 341 | 426 | 392 | 387 | 353 | 348 | 344 | 339 | 335 | 331 | 326 | 335 |
| 6 | ... | 325 | 380 | 407 | 402 | 367 | 362 | 358 | 353 | 349 | 315 | 310 | 314 |
| 7 | 374 | 490 | 575 | 571 | 597 | 593 | 527 | 553 | 488 | 454 | 450 | 475 | 412 |
| 8 | 599 | 655 | 801 | 857 | 912 | 938 | 843 | 839 | 292 | 529 | 103 | (—)717 | 540 |
| 9 | 403 | 639 | 845 | 1232 | 1047 | 952 | 797 | 703 | 728 | 724 | 720 | 414 | 106 |
| 10 | 899 | 1015 | 1100 | 1187 | 1182 | 1238 | 1263 | 1018 | 983 | 1009 | 704 | 850 | 642 |
| 11 | 1184 | 1360 | 1295 | 1111 | 1136 | 1282 | 1187 | 1093 | 1088 | 1114 | 809 | 894 | 960 |
| 12 | ... | ... | ... | ... | ... | ... | ... | ... | ... | ... | ... | ... | 975 |
| 13 | ... | ... | ... | ... | ... | ... | ... | ... | ... | ... | ... | ... | ... |
| 14 | 1194 | 1219 | 1214 | 1361 | 1325 | 1291 | 1316 | 1162 | 1307 | 1213 | 1209 | 1205 | 1251 |
| 15 | 1293 | 1291 | 1318 | 1316 | 1313 | 1281 | 1398 | 1366 | 1363 | 1391 | 1358 | 1356 | 1303 |
| 16 | 1114 | 1263 | 1380 | 1318 | 1345 | 1493 | 1521 | 1488 | 1666 | 1423 | 1390 | 1448 | 1387 |
| 17 | ... | ... | ... | ... | ... | ... | ... | ... | ... | ... | ... | ... | 1475 |
| 18 | 1781 | 1658 | 1746 | 1894 | 1861 | 1919 | 2007 | 2005 | 1911 | 1879 | 1816 | 1874 | 1862 |
| 19 | 1602 | 1721 | 1778 | 1866 | 1863 | 1831 | 1828 | 1826 | 1793 | 1821 | 1788 | 1786 | 1809 |
| 20 | 1906 | 1934 | 1931 | 1929 | 1896 | 1894 | 1891 | 1889 | 1886 | 1884 | 1881 | 1879 | 1870 |
| 21 | 1547 | 1605 | 1662 | 1690 | 1627 | 1625 | 1592 | 1620 | 1587 | 1615 | 1642 | 1640 | 1721 |
| 22 | 1579 | 1757 | 1935 | 2023 | 1960 | 2109 | 1955 | 1893 | 1860 | 1828 | 1765 | 1763 | 1708 |
| 23 | 1733 | 1761 | 1818 | 1816 | 1843 | 1841 | 1778 | 1746 | 1713 | 1711 | 1647 | 1675 | 1659 |
| 24 | 1041 | 1009 | 1036 | 1004 | 1062 | 999 | 996 | 934 | 841 | 869 | 836 | 834 | 1261 |
| 25 | 713 | 711 | 738 | 766 | 763 | 791 | 758 | 786 | 753 | 691 | 718 | 716 | 745 |
| 26 | ... | 833 | 770 | 798 | 735 | 763 | 700 | 698 | 665 | 693 | 660 | 658 | 691 |
| 27 | 807 | 805 | 802 | 830 | 888 | 886 | 913 | 911 | 908 | 876 | 842 | 810 | 753 |
| 28 | 1020 | 988 | 895 | 893 | 920 | 918 | 885 | 883 | 820 | 818 | 815 | 813 | 882 |
| 29 | 902 | 870 | 897 | 895 | 862 | 800 | 797 | 795 | 792 | 790 | 787 | 815 | 804 |
| 30 | 693 | 661 | 688 | 717 | 714 | 681 | 678 | 676 | 704 | 702 | 699 | 697 | 710 |
| Mean. | 961 | 982 | 1030 | 1059 | 1049 | 1051 | 1031 | 1003 | 978 | 977 | 958 | 900 | 927 |

TABLE VII—Continued—*Ver*

The figures given in the table are millionths of a dyne, which,

.579000+

| Day. | A. M. | | | | | | | | | | | |
|---|---|---|---|---|---|---|---|---|---|---|---|---|
| | 1. | 2. | 3. | 4. | 5. | 6. | 7. | 8. | 9. | 10. | 11. | 12. |
| 1 | 674 | 672 | 669 | 667 | 664 | 662 | 659 | 626 | 654 | 591 | 498 | 496 |
| 2 | 134 | 132 | 099 | 097 | 094 | 092 | 119 | 117 | 114 | 082 | 019 | (—)044 |
| 3 | (—)014 | (—)016 | 011 | 099 | 126 | 154 | 181 | 179 | 176 | 144 | 141 | 139 |
| 4 | 199 | 197 | 163 | 161 | 158 | 156 | 184 | 182 | 148 | 116 | 083 | 021 |
| 5 | 1495 | 1463 | 1490 | 1488 | 1485 | 1513 | 1540 | 1538 | 1505 | 1383 | 1320 | 1378 |
| 6 | 1226 | 1315 | 1342 | 1249 | 1277 | 1455 | 1452 | 1480 | 1477 | 1445 | 1442 | 1470 |
| 7 | 1530 | 1528 | 1464 | 1553 | 1640 | 1668 | 1695 | 1663 | 1720 | 1718 | 1715 | 1713 |
| 8 | 1712 | 1710 | 1707 | 1705 | 1763 | 1761 | 1758 | 1786 | 1783 | 1781 | 1778 | 1806 |
| 9 | 1805 | 1802 | 1800 | 1827 | 1825 | 1852 | 1880 | 1877 | 1845 | 1812 | 1810 | 1748 |
| 10 | 1386 | 1323 | 1351 | 1318 | 1376 | 1374 | 1401 | 1369 | 1396 | 1364 | 1270 | 1268 |
| 11 | 1930 | 1898 | 1895 | 1923 | 1920 | 1948 | 1975 | 2003 | 2000 | 1938 | 1875 | 1842 |
| 12 | 1691 | 1659 | 1656 | 1654 | 1651 | 1649 | 1646 | 1674 | 1701 | 1639 | 1606 | 1544 |
| 13 | 1392 | 1420 | 1387 | 1355 | 1382 | 1350 | 1347 | 1345 | 1312 | 1250 | 1187 | 1185 |
| 14 | 1364 | 1332 | 1359 | 1357 | 1384 | 1413 | 1440 | 1468 | 1465 | 1463 | 1400 | 1367 |
| 15 | 1698 | 1696 | 1693 | 1691 | 1688 | 1716 | 1713 | 1711 | 1708 | 1676 | 1552 | 1550 |
| 16 | 1730 | 1668 | 1695 | 1753 | 1750 | 1748 | 1775 | 1773 | 1740 | 1708 | 1645 | 1643 |
| 17 | 1672 | 1670 | 1667 | 1665 | 1662 | 1690 | 1717 | 1685 | 1682 | 1650 | 1617 | 1615 |
| 18 | 1495 | 1523 | 1520 | 1518 | 1485 | 1483 | 1510 | 1508 | 1475 | 1442 | 1379 | 1347 |
| 19 | 1286 | 1314 | 1281 | 1249 | 1246 | 1214 | 1241 | 1209 | 1206 | 1144 | 1141 | 1169 |
| 20 | 1108 | 1136 | 1073 | 1010 | 887 | 885 | 882 | 940 | 967 | 935 | 902 | 930 |
| 21 | 719 | 626 | 533 | 591 | 709 | 706 | 704 | 761 | 699 | 696 | 633 | 661 |
| 22 | 782 | 780 | 807 | 805 | 832 | 890 | 827 | 855 | 882 | 880 | 877 | 845 |
| 23 | 965 | 963 | 960 | 988 | 955 | 983 | 980 | 978 | 1035 | 1033 | 970 | 1057 |
| 24 | 1569 | 1627 | 1654 | 1682 | 1709 | 1738 | 1735 | 1763 | 1730 | 1697 | 1664 | 1662 |
| 25 | 1782 | 1780 | 1747 | 1745 | 1772 | 1740 | 1737 | 1735 | 1702 | 1670 | 1636 | 1574 |
| 26 | 1454 | 1452 | 1419 | 1417 | 1444 | 1442 | 1409 | 1377 | 1314 | 1281 | 1278 | 1276 |
| 27 | 1245 | 1243 | 1240 | 1238 | 1205 | 1173 | 1200 | 1138 | 1105 | 1043 | 1040 | 1008 |
| 28 | 885 | 853 | 880 | 878 | 875 | 873 | 870 | 899 | 896 | 863 | 800 | 798 |
| 29 | 950 | 948 | 945 | 943 | 970 | 998 | 965 | 963 | 960 | 898 | 865 | 893 |
| 30 | 1042 | 1040 | 1037 | 1035 | 1062 | 1060 | 1057 | 1086 | 1083 | 1050 | 1047 | 1045 |
| 31 | 864 | 862 | 861 | 857 | 854 | 851 | 849 | 846 | 814 | 721 | 508 | 566 |
| Mean . | 1218 | 1213 | 1203 | 1210 | 1221 | 1233 | 1240 | 1243 | 1235 | 1197 | 1152 | 1149 |

*tical force, October,* 1889.

added to .579000 dyne, give the vertical force in C. G. S. units.

.579000 +

| Day. | P. M. | | | | | | | | | | | | Mean. |
|---|---|---|---|---|---|---|---|---|---|---|---|---|---|
| | 1. | 2. | 3. | 4. | 5. | 6. | 7. | 8. | 9. | 10. | 11. | 12. | |
| 1 | 283 | 311 | 218 | 186 | 243 | 211 | 148 | 176 | 143 | 141 | 138 | 136 | 411 |
| 2 | (—)046 | (—)017 | 040 | 068 | 065 | 093 | 090 | 088 | 115 | 113 | 110 | 108 | 078 |
| 3 | 167 | 165 | 192 | 221 | 187 | 185 | 182 | 211 | 208 | 206 | 203 | 201 | 152 |
| 4 | 1313 | 1311 | 1399 | 1427 | 1484 | 1542 | 1539 | 1537 | 1534 | 1532 | 1529 | 1497 | 809 |
| 5 | 1466 | 1524 | 1521 | 1579 | 1727 | 1725 | 1752 | 1780 | 1566 | 1263 | 1200 | 1198 | 1496 |
| 6 | 1498 | 1526 | 1554 | 1552 | 1639 | 1607 | 1574 | 1572 | 1539 | 1476 | 1443 | 1461 | 1461 |
| 7 | 1561 | 1649 | 1676 | 1704 | 1671 | 1699 | 1696 | 1694 | 1661 | 1689 | 1536 | 1594 | 1643 |
| 8 | 1834 | 1862 | 1859 | 1887 | 1824 | 1822 | 1819 | 1817 | 1814 | 1751 | 1779 | 1807 | 1789 |
| 9 | 1655 | 1623 | 1590 | 1588 | 1525 | 1553 | 1520 | 1458 | 1455 | 1423 | 1420 | 1388 | 1670 |
| 10 | 1989 | 2047 | 2044 | 2012 | 1949 | 1977 | 1974 | 1972 | 1969 | 1967 | 1964 | 1932 | 1662 |
| 11 | 1961 | 1959 | 1926 | 1924 | 1830 | 1828 | 1795 | 1763 | 1730 | 1758 | 1695 | 1693 | 1875 |
| 12 | 1512 | 1510 | 1537 | 1565 | 1532 | 1530 | 1466 | 1464 | 1461 | 1459 | 1396 | 1394 | 1566 |
| 13 | 1243 | 1181 | 1178 | 1206 | 1263 | 1291 | 1348 | 1316 | 1313 | 1281 | 1338 | 1366 | 1302 |
| 14 | 1396 | 1454 | 1481 | 1539 | 1566 | 1594 | 1591 | 1619 | 1616 | 1614 | 1672 | 1670 | 1484 |
| 15 | 1669 | 1667 | 1694 | 1812 | 1899 | 1867 | 1925 | 1892 | 1859 | 1827 | 1824 | 1822 | 1744 |
| 16 | 1611 | 1639 | 1636 | 1694 | 1691 | 1689 | 1686 | 1684 | 1681 | 1679 | 1676 | 1644 | 1693 |
| 17 | 1673 | 1701 | 1638 | 1636 | 1633 | 1571 | 1598 | 1596 | 1593 | 1561 | 1529 | 1527 | 1635 |
| 18 | 1315 | 1343 | 1400 | 1489 | 1486 | 1454 | 1390 | 1358 | 1325 | 1353 | 1260 | 1318 | 1424 |
| 19 | 1197 | 1195 | 1192 | 1190 | 1187 | 1185 | 1212 | 1270 | 1207 | 1085 | 1082 | 1080 | 1199 |
| 20 | 898 | 926 | 953 | 981 | 978 | 946 | 943 | 911 | 848 | 906 | 843 | 781 | 945 |
| 21 | 750 | 808 | 835 | 863 | 920 | 918 | 885 | 913 | 820 | 848 | 756 | 784 | 756 |
| 22 | 813 | 811 | 868 | 896 | 954 | 982 | 979 | 1007 | 1004 | 972 | 969 | 967 | 887 |
| 23 | 1177 | 1205 | 1232 | 1290 | 1317 | 1405 | 1432 | 1491 | 1578 | 1606 | 1603 | 1601 | 1200 |
| 24 | 1751 | 1809 | 1806 | 1804 | 1801 | 1799 | 1796 | 1794 | 1791 | 1789 | 1786 | 1784 | 1739 |
| 25 | 1572 | 1570 | 1597 | 1565 | 1532 | 1500 | 1527 | 1525 | 1552 | 1550 | 1517 | 1485 | 1630 |
| 26 | 1305 | 1303 | 1300 | 1298 | 1295 | 1262 | 1290 | 1288 | 1285 | 1283 | 1280 | 1278 | 1335 |
| 27 | 1006 | 974 | 941 | 908 | 905 | 903 | 900 | 929 | 926 | 924 | 921 | 919 | 1043 |
| 28 | 796 | 854 | 882 | 910 | 907 | 935 | 932 | 900 | 927 | 925 | 922 | 920 | 883 |
| 29 | 861 | 889 | 916 | 914 | 941 | 939 | 966 | 1024 | 1051 | 1019 | 1046 | 1044 | 954 |
| 30 | 893 | 891 | 888 | 886 | 823 | 821 | 818 | 846 | 873 | 871 | 868 | 866 | 958 |
| 31 | 594 | 682 | 710 | 738 | 674 | 703 | 700 | 698 | 695 | 693 | 690 | 688 | 738 |
| Mean. | 1217 | 1238 | 1248 | 1278 | 1273 | 1275 | 1473 | 1277 | 1263 | 1245 | 1232 | 1224 | 1231 |

9230——6

TABLE VII—Continued—*Ver*

The figures given in the table are millionths of a dyne, which,

.578000 +

| Day. | A. M. | | | | | | | | | | | |
|---|---|---|---|---|---|---|---|---|---|---|---|---|
| | 1. | 2. | 3. | 4. | 5. | 6. | 7. | 8. | 9. | 10. | 11. | 12. |
| 1 | 1635 | 1392 | 1148 | 1267 | 1324 | 1412 | 1439 | 1467 | 1464 | 1492 | 1670 | 1668 |
| 2 | 1456 | 1394 | 1331 | 1389 | 1386 | 1444 | 1411 | 1409 | 1346 | 1344 | 1281 | 1249 |
| 3 | 1189 | 1428 | 1094 | 1122 | 1149 | 1177 | 1174 | 1112 | 1050 | 1016 | 923 | 921 |
| 4 | 1041 | 1039 | 1126 | 1154 | 1211 | 1239 | 1236 | 1234 | 1201 | 1199 | 1226 | 1224 |
| 5 | 1314 | 1312 | 1309 | 1367 | 1394 | 1392 | 1389 | 1417 | 1414 | 1382 | 1349 | 1377 |
| 6 | 1647 | 1675 | 1672 | 1700 | 1697 | 1695 | 1722 | 1690 | 1687 | 1655 | 1652 | 1710 |
| 7 | 1831 | 1829 | 1826 | 1824 | 1851 | 1879 | 1876 | 1874 | 1871 | 1839 | 1806 | 1834 |
| 8 | 1833 | 1831 | 1828 | 1826 | 1823 | 1821 | 1788 | 1786 | 1753 | 1721 | 1718 | 1776 |
| 9 | 1746 | 1714 | 1711 | 1709 | 1706 | 1674 | 1671 | 1699 | 1666 | 1604 | 1570 | 1598 |
| 10 | 1658 | 1656 | 1623 | 1621 | 1618 | 1616 | 1613 | 1641 | 1578 | 1515 | 1452 | 1450 |
| 11 | 1389 | 1387 | 1384 | 1412 | 1409 | 1377 | 1344 | 1402 | 1369 | 1367 | 1364 | 1362 |
| 12 | 1512 | 1480 | 1477 | 1444 | 1441 | 1409 | 1376 | 1344 | 1311 | 1279 | 1276 | 1302 |
| 13 | 1635 | 1663 | 1630 | 1628 | 1685 | 1683 | 1680 | 1678 | 1645 | 1643 | 1640 | 1638 |
| 14 | 464 | 462 | 398 | 396 | 393 | 391 | 388 | 356 | 353 | 321 | 318 | 256 |
| 15 | 1158 | 1186 | 1243 | 1302 | 1299 | 1387 | 1384 | 1442 | 1439 | 1437 | 1374 | 1402 |
| 16 | 1220 | 1098 | 1125 | 1213 | 1301 | 1299 | 1326 | 1324 | 1321 | 1319 | 1316 | 1314 |
| 17 | 1404 | 1432 | 1369 | 1337 | 1334 | 1391 | 1449 | 1477 | 1474 | 1472 | 1469 | 1467 |
| 18 | 1166 | 1156 | 1176 | 1196 | 1306 | 1326 | 1316 | 1306 | 1296 | 1286 | 1276 | 1296 |
| 19 | 897 | 856 | 846 | 776 | 766 | 726 | 656 | 676 | 636 | 566 | 586 | 606 |
| 20 | 388 | 378 | 368 | 358 | 378 | 368 | 328 | 318 | 308 | 358 | 619 | 579 |
| 21 | 1594 | 1584 | 1574 | 1534 | 1524 | 1514 | 1534 | 1524 | 1454 | 1384 | 1705 | 1996 |
| 22 | 2981 | 2971 | 2931 | 2861 | 2881 | 2841 | 2800 | 2760 | 2690 | 2560 | 2580 | 2660 |
| 23 | 2591 | 2581 | 2601 | 2652 | 2642 | 2602 | 2622 | 2612 | 2541 | 2441 | 2250 | 2240 |
| 24 | 2683 | 2734 | 2724 | 2714 | 2734 | 2724 | 2714 | 2704 | 2664 | 2503 | 2373 | 2393 |
| 25 | 2867 | 2797 | 2666 | 2626 | 2676 | 2666 | 2656 | 2586 | 2365 | 2325 | 2255 | 2305 |
| 26 | 2387 | 2317 | 2307 | 2297 | 2287 | 2307 | 2146 | 2166 | 2217 | 2207 | 2257 | 2428 |
| 27 | 2569 | 2650 | 2760 | 2780 | 2770 | 2760 | 2780 | 2680 | 2519 | 2419 | 2319 | 2309 |
| 28 | 2391 | 2320 | 2220 | 2180 | 2110 | 2010 | 1909 | 1899 | 1769 | 1699 | 1598 | 1709 |
| 29 | 1610 | 1510 | 1891 | 1971 | 1991 | 2041 | 2152 | 2142 | 2192 | 2152 | 2112 | 2192 |
| 30 | 2215 | 1934 | 1894 | 1884 | 1924 | 1894 | 1884 | 1994 | 1984 | 1914 | 1844 | 2075 |
| Mean | 1682 | 1659 | 1642 | 1651 | 1667 | 1669 | 1659 | 1657 | 1619 | 1580 | 1539 | 1611 |

*tical force, November,* 1889.

added to .578000 dyne, give the vertical force in C. G. S. units.

.578000 +

| Day. | P. M. | | | | | | | | | | | | Mean. |
| | 1. | 2. | 3. | 4. | 5. | 6. | 7. | 8. | 9. | 10. | 11. | 12. | |
|---|---|---|---|---|---|---|---|---|---|---|---|---|---|
| 1 | 1696 | 1664 | 1721 | 1689 | 1867 | 1744 | 1802 | 1769 | 1797 | 1704 | 1581 | 1489 | 1579 |
| 2 | 1367 | 1425 | 1452 | 1450 | 1447 | 1445 | 1442 | 1440 | 1437 | 1435 | 1373 | 1281 | 1393 |
| 3 | 979 | 1007 | 1034 | 1062 | 1059 | 1057 | 1085 | 1083 | 1050 | 957 | 1014 | 1042 | 1074 |
| 4 | 1282 | 1341 | 1368 | 1366 | 1333 | 1361 | 1358 | 1356 | 1353 | 1321 | 1348 | 1316 | 1260 |
| 5 | 1435 | 1493 | 1490 | 1488 | 1455 | 1483 | 1541 | 1659 | 1656 | 1683 | 1651 | 1649 | 1462 |
| 6 | 1738 | 1796 | 1793 | 1761 | 1788 | 1786 | 1783 | 1781 | 1778 | 1807 | 1774 | 1803 | 1733 |
| 7 | 2012 | 2040 | 1977 | 1975 | 1942 | 1940 | 1907 | 1935 | 1902 | 1899 | 1867 | 1865 | 1892 |
| 8 | 1774 | 1772 | 1769 | 1767 | 1734 | 1732 | 1759 | 1757 | 1754 | 1752 | 1750 | 1748 | 1774 |
| 9 | 1747 | 1745 | 1742 | 1740 | 1677 | 1675 | 1732 | 1760 | 1757 | 1755 | 1662 | 1660 | 1697 |
| 10 | 1448 | 1476 | 1473 | 1441 | 1438 | 1436 | 1433 | 1461 | 1428 | 1456 | 1423 | 1391 | 1514 |
| 11 | 1480 | 1448 | 1475 | 1473 | 1410 | 1438 | 1465 | 1524 | 1551 | 1549 | 1546 | 1514 | 1435 |
| 12 | 1844 | 1872 | 1869 | 1837 | 1804 | 1742 | 1739 | 1707 | 1703 | 1671 | 1669 | 1667 | 1574 |
| 13 | 733 | 731 | 728 | 666 | 633 | 601 | 568 | 506 | 503 | 501 | 498 | 496 | 1126 |
| 14 | 346 | 372 | 369 | 337 | 575 | 663 | 720 | 869 | 926 | 1014 | 1072 | 1130 | 537 |
| 15 | 1881 | 1699 | 1485 | 1393 | 1330 | 1328 | 1295 | 1263 | 1260 | 1228 | 1224 | 1223 | 1361 |
| 16 | 1342 | 1400 | 1397 | 1365 | 1302 | 1300 | 1327 | 1325 | 1352 | 1380 | 1348 | 1346 | 1307 |
| 17 | 1225 | 1223 | 1250 | 1278 | 1335 | 1303 | 1330 | 1388 | 1295 | 1323 | 1380 | 1378 | 1366 |
| 18 | 1226 | 1186 | 1176 | 1136 | 1066 | 1026 | 1016 | 1006 | 966 | 956 | 946 | 936 | 1156 |
| 19 | 506 | 467 | 396 | 326 | 286 | 276 | 296 | 316 | 366 | 356 | 376 | 366 | 539 |
| 20 | 1803 | 1764 | 1724 | 1653 | 1643 | 1633 | 1653 | 1643 | 1603 | 1623 | 1613 | 1603 | 1029 |
| 21 | 3070 | 3150 | 3140 | 3070 | 3030 | 3050 | 3010 | 3030 | 3110 | 3160 | 3120 | 3020 | 2328 |
| 22 | 2771 | 2851 | 2781 | 2680 | 2670 | 2630 | 2620 | 2610 | 2570 | 2560 | 2580 | 2570 | 2725 |
| 23 | 2682 | 2732 | 2752 | 2742 | 2702 | 2752 | 2772 | 2793 | 2783 | 2773 | 2733 | 2723 | 2655 |
| 24 | 2925 | 3006 | 3056 | 3016 | 3036 | 2996 | 2986 | 2976 | 2996 | 2956 | 2976 | 2906 | 2812 |
| 25 | 2235 | 2285 | 2335 | 2295 | 2255 | 2305 | 2386 | 2436 | 2486 | 2446 | 2406 | 2396 | 2461 |
| 26 | 2899 | 3130 | 3481 | 3381 | 3461 | 3692 | 3953 | 4153 | 3993 | 3742 | 3582 | 3090 | 2912 |
| 27 | 2299 | 2349 | 2580 | 2600 | 2650 | 2941 | 3142 | 3102 | 2881 | 2841 | 2590 | 2520 | 2659 |
| 28 | 2722 | 2291 | 2401 | 2210 | 2200 | 2371 | 2421 | 2351 | 2341 | 2301 | 2261 | 1980 | 2157 |
| 29 | 2393 | 2233 | 2283 | 2273 | 2294 | 2374 | 2394 | 2384 | 2407 | 2424 | 2234 | 2254 | 2163 |
| 30 | 2125 | 2175 | 2285 | 2275 | 2205 | 2195 | 2185 | 2205 | 2165 | 2185 | 2205 | 2165 | 2075 |
| Mean | 1800 | 1804 | 1826 | 1791 | 1787 | 1809 | 1837 | 1853 | 1835 | 1825 | 1793 | 1751 | 1723 |

TABLE VII—Continued—*Ver*

The figures given in the table are millionths of a dyne, which,

.57900c +

| Day. | A. M. | | | | | | | | | | | |
|---|---|---|---|---|---|---|---|---|---|---|---|---|
| | 1. | 2. | 3. | 4. | 5. | 6. | 7. | 8. | 9. | 10. | 11. | 12. |
| 1 | 1167 | 1127 | 967 | 927 | 947 | 967 | 987 | 947 | 967 | 927 | 917 | 907 |
| 2 | 928 | 968 | 948 | 898 | 908 | 948 | 898 | 878 | 858 | 868 | 968 | 918 |
| 3 | 1077 | 1087 | 1157 | 1227 | 1237 | 1277 | 1257 | 1267 | 1277 | 1257 | 1207 | 1217 |
| 4 | 1194 | 1204 | 1244 | 1254 | 1264 | 1274 | 1254 | 1264 | 1274 | 1314 | 1324 | 1304 |
| 5 | 1462 | 1442 | 1482 | 1523 | 1593 | 1603 | 1613 | 1683 | 1783 | 1733 | 1623 | 1663 |
| 6 | 1852 | 1893 | 1902 | 1912 | 1963 | 1842 | 1762 | 1681 | 1511 | 1430 | 1350 | 1330 |
| 7 | 1639 | 1619 | 1659 | 1669 | 1679 | 1689 | 1730 | 1740 | 1719 | 1699 | 1559 | 1629 |
| 8 | 1638 | 1678 | 1628 | 1607 | 1617 | 1597 | 1637 | 1647 | 1627 | 1547 | 1467 | 1417 |
| 9 | 1545 | 1585 | 1595 | 1605 | 1615 | 1595 | 1575 | 1404 | 1354 | 1273 | 1163 | 1173 |
| 10 | 911 | 891 | 871 | 850 | 1161 | 1141 | 790 | 770 | 750 | 579 | 589 | 629 |
| 11 | 1210 | 1220 | 1260 | 1270 | 1280 | 1290 | 1240 | 1250 | 1230 | 1210 | 1069 | 1139 |
| 12 | 1268 | 1278 | 1288 | 1238 | 1248 | 1228 | 1208 | 1218 | 1228 | 1238 | 1158 | 1107 |
| 13 | 1537 | 1517 | 1618 | 1598 | 1547 | 1587 | 1628 | 1607 | 1527 | 1507 | 1427 | 1437 |
| 14 | 1264 | 1093 | 1314 | 1324 | 1334 | 1344 | 1354 | 1364 | 1434 | 1384 | 1334 | 1314 |
| 15 | 961 | 941 | 921 | 961 | 971 | 1011 | 1021 | 1091 | 1101 | 1021 | 941 | 1011 |
| 16 | 1320 | 1330 | 1370 | 1410 | 1390 | 1400 | 1410 | 1450 | 1430 | 1230 | 1300 | 1460 |
| 17 | 1890 | 1870 | 1850 | 1710 | 1660 | 1640 | 1619 | 1659 | 1459 | 1439 | 1449 | 1489 |
| 18 | 986 | 966 | 946 | 925 | 905 | 885 | 805 | 815 | 795 | 775 | 694 | 795 |
| 19 | 695 | 685 | 675 | 695 | 655 | 645 | 605 | 595 | 525 | 395 | 385 | 465 |
| 20 | 1179 | 1199 | 1189 | 1149 | 1109 | 1099 | 1089 | 1019 | 888 | 818 | 748 | 798 |
| 21 | 639 | 599 | 529 | 488 | 508 | 438 | 428 | 418 | 378 | 338 | 358 | 378 |
| 22 | 1182 | 1172 | 1162 | 1182 | 1202 | 1192 | 1182 | 1172 | 1102 | 971 | 961 | 1012 |
| 23 | 1244 | 1264 | 1254 | 1214 | 1295 | 1315 | 1335 | 1355 | 1285 | 1064 | 1144 | 1255 |
| 24 | 1156 | 1116 | 1106 | 1095 | 1116 | 1136 | 1126 | 1146 | 1046 | 945 | 1116 | 1106 |
| 25 | 796 | 816 | 776 | 676 | 666 | 686 | 646 | 696 | 566 | 466 | 486 | 506 |
| 26 | 1250 | 1210 | 1200 | 1190 | 1150 | 1110 | 1130 | 1150 | 1049 | 889 | 819 | 809 |
| 27 | 619 | 579 | 569 | 711 | 821 | 841 | 981 | 1002 | 811 | 681 | 641 | 661 |
| 28 | 862 | 882 | 872 | 862 | 852 | 842 | 832 | 852 | 812 | 501 | 642 | 782 |
| 29 | 1345 | 1274 | 1294 | 1284 | 1274 | 1234 | 1284 | 1274 | 1234 | 1164 | 1184 | 1204 |
| 30 | 926 | 856 | 755 | 775 | 735 | 695 | 715 | 765 | 725 | 565 | 815 | 866 |
| 31 | ... | ... | ... | ... | ... | ... | ... | ... | ... | ... | ... | ... |
| Mean. | 1191 | 1179 | 1180 | 1174 | 1190 | 1185 | 1171 | 1173 | 1131 | 1041 | 1028 | 1059 |

*tical force, December,* 1889.

added to .579000 dyne, give the vertical force in C. G. S. units.

.579000 +

| Day. | P. M. | | | | | | | | | | | | Mean. |
|---|---|---|---|---|---|---|---|---|---|---|---|---|---|
| | 1. | 2. | 3. | 4. | 5. | 6. | 7. | 8. | 9. | 10. | 11. | 12. | |
| 1 | 1077 | 1097 | 997 | 1077 | 1007 | 967 | 987 | 1007 | 997 | 987 | 977 | 937 | 995 |
| 2 | 988 | 878 | 888 | 868 | 908 | 918 | 928 | 938 | 978 | 988 | 998 | 1008 | 928 |
| 3 | 1498 | 1387 | 1367 | 1377 | 1417 | 1367 | 1287 | 1207 | 1217 | 1227 | 1207 | 1216 | 1267 |
| 4 | 1194 | 1234 | 1304 | 1344 | 1354 | 1394 | 1374 | 1384 | 1394 | 1434 | 1444 | 1453 | 1311 |
| 5 | 1853 | 1893 | 1903 | 1853 | 1863 | 1873 | 1013 | 1893 | 1873 | 1883 | 1893 | 1873 | 1740 |
| 6 | 1490 | 1591 | 1601 | 1581 | 1561 | 1571 | 1581 | 1561 | 1691 | 1671 | 1651 | 1631 | 1650 |
| 7 | 1699 | 1709 | 1719 | 1699 | 1649 | 1689 | 1759 | 1799 | 1809 | 1789 | 1799 | 1538 | 1695 |
| 8 | 1577 | 1706 | 1596 | 1486 | 1496 | 1506 | 1515 | 1526 | 1596 | 1576 | 1556 | 1566 | 1575 |
| 9 | 852 | 982 | 992 | 1002 | 892 | 932 | 912 | 892 | 902 | 912 | 922 | 932 | 1192 |
| 10 | 1302 | 1101 | 1051 | 1031 | 1041 | 1141 | 1151 | 1131 | 1171 | 1181 | 1191 | 1171 | 983 |
| 11 | 1300 | 1310 | 1230 | 1179 | 1129 | 1169 | 1209 | 1219 | 1229 | 1299 | 1309 | 1259 | 1229 |
| 12 | 1298 | 1308 | 1288 | 1298 | 1278 | 1288 | 1298 | 1338 | 1408 | 1478 | 1518 | 1528 | 1293 |
| 13 | 1447 | 1487 | 1437 | 1417 | 1366 | 1346 | 1326 | 1336 | 1376 | 1386 | 1396 | 1376 | 1468 |
| 14 | 721 | 731 | 711 | 811 | 821 | 831 | 841 | 881 | 891 | 901 | 911 | 921 | 1076 |
| 15 | 1021 | 1031 | 981 | 991 | 1031 | 1071 | 1111 | 1151 | 1191 | 1331 | 1401 | 1441 | 1071 |
| 16 | 1832 | 1812 | 1792 | 1771 | 1721 | 1791 | 1922 | 1962 | 1912 | 1892 | 1902 | 1882 | 1612 |
| 17 | 1107 | 1147 | 1007 | 987 | 967 | 977 | 1017 | 1027 | 1037 | 1047 | 1027 | 1007 | 1367 |
| 18 | 865 | 795 | 664 | 594 | 584 | 604 | 594 | 644 | 694 | 684 | 584 | 544 | 756 |
| 19 | 1358 | 1348 | 1308 | 1328 | 1318 | 1308 | 1328 | 1318 | 1368 | 1328 | 1258 | 1218 | 950 |
| 20 | 878 | 868 | 798 | 788 | 778 | 768 | 788 | 808 | 738 | 728 | 718 | 678 | 901 |
| 21 | 1030 | 960 | 920 | 1000 | 1020 | 1071 | 1091 | 1111 | 1131 | 1181 | 1261 | 1221 | 771 |
| 22 | 1664 | 1353 | 1343 | 1424 | 1444 | 1404 | 1454 | 1444 | 1404 | 1424 | 1294 | 1163 | 1263 |
| 23 | 1185 | 1024 | 1044 | 1094 | 1084 | 1074 | 1125 | 1145 | 1195 | 1185 | 1205 | 1165 | 1194 |
| 24 | 1006 | 845 | 835 | 765 | 755 | 745 | 795 | 846 | 775 | 795 | 846 | 806 | 959 |
| 25 | 1549 | 1449 | 1379 | 1369 | 1329 | 1289 | 1279 | 1329 | 1349 | 1339 | 1299 | 1259 | 1000 |
| 26 | 738 | 849 | 779 | 829 | 729 | 719 | 799 | 819 | 989 | 949 | 909 | 779 | 952 |
| 27 | 711 | 761 | 751 | 741 | 701 | 691 | 741 | 761 | 721 | 802 | 792 | 812 | 746 |
| 28 | 1645 | 1635 | 1565 | 1525 | 1424 | 1414 | 1404 | 1394 | 1445 | 1435 | 1304 | 1324 | 1129 |
| 29 | 1285 | 1305 | 1265 | 1285 | 1154 | 1114 | 1074 | 1064 | 1054 | 984 | 944 | 934 | 1188 |
| 30 | ... | ... | ... | ... | ... | ... | ... | ... | ... | ... | ... | ... | 766 |
| 31 | ... | ... | ... | ... | ... | ... | ... | ... | ... | ... | ... | ... | |
| Mean. | 1247 | 1227 | 1190 | 1190 | 1166 | 1174 | 1193 | 1205 | 1225 | 1235 | 1225 | 1195 | 1174 |

## TABLE VIII.—*Disturbances in declination.*

In the columns headed No. are given the number of traces of declination for that month and hour which differ 2 minutes or more from the mean declination, as shown by the composite curve, for that month and hour. The columns headed Value give the sum of these differences in minutes of arc.

1888.

| Hour | January | | | | | | February | | | | | | March | | | | | |
|---|---|---|---|---|---|---|---|---|---|---|---|---|---|---|---|---|---|---|
| | Easterly | | Westerly | | Total | | Easterly | | Westerly | | Total | | Easterly | | Westerly | | Total | |
| | No. | Value. | No. | Value. | No. | Value. | No. | Value. | No. | Value. | No. | Value. | No. | Value. | No. | Value. | No. | Value. |
| A. M. | | ′ | | ′ | | ′ | | ′ | | ′ | | ′ | | ′ | | ′ | | ′ |
| 1 | 1 | 5.3 | 0 | 0 | 1 | 5.3 | 0 | 0 | 1 | 7.0 | 1 | 7.0 | 4 | 19.5 | 0 | 0 | 4 | 19.5 |
| 2 | 4 | 13.8 | 3 | 7.0 | 7 | 20.8 | 0 | 0 | 1 | 2.6 | 1 | 2.6 | 4 | 11.3 | 3 | 7.7 | 7 | 19.0 |
| 3 | 3 | 13.9 | 1 | 3.0 | 4 | 16.9 | 3 | 6.0 | 2 | 5.0 | 5 | 11.0 | 2 | 5.3 | 2 | 13.0 | 4 | 18.3 |
| 4 | 2 | 13.5 | 1 | 4.0 | 3 | 17.5 | 3 | 6.0 | 1 | 4.0 | 4 | 10 0 | 0 | 0 | 2 | 5.9 | 2 | 5.9 |
| 5 | 2 | 11.0 | 1 | 2.0 | 3 | 13.0 | 1 | 4.2 | 0 | 0 | 1 | 4.2 | 3 | 9.5 | 0 | 0 | 3 | 9.5 |
| 6 | 1 | 6.0 | 1 | 18.0 | 2 | 24.0 | 1 | 2.5 | 1 | 5.0 | 2 | 7.5 | 0 | 0 | 2 | 6.7 | 2 | 6.7 |
| 7 | 1 | 3.0 | 2 | 22.5 | 3 | 25.5 | 3 | 7.0 | 1 | 3.0 | 4 | 10.0 | 0 | 0 | 3 | 13.7 | 3 | 13.7 |
| 8 | 3 | 9.7 | 3 | 24.5 | 6 | 34.2 | 0 | 0 | 0 | 0 | 0 | 0 | 1 | 3.0 | 2 | 11.0 | 3 | 14.0 |
| 9 | 1 | 3.0 | 3 | 34.7 | 4 | 37.7 | 0 | 0 | 3 | 9.0 | 3 | 9.0 | 0 | 0 | 3 | 17.5 | 3 | 17.5 |
| 10 | 0 | 0 | 2 | 18.9 | 2 | 18.9 | 2 | 5.0 | 2 | 5.0 | 4 | 10.0 | 1 | 3.5 | 3 | 11.5 | 4 | 15.0 |
| 11 | 0 | 0 | 1 | 4.0 | 1 | 4.0 | 0 | 0 | 0 | 0 | 0 | 0 | 1 | 4.0 | 1 | 4.0 | 2 | 8.0 |
| 12 | 1 | 4.2 | 1 | 3.0 | 2 | 7.2 | 0 | 0 | 0 | 0 | 0 | 0 | 0 | 0 | 1 | 4.0 | 1 | 4.0 |
| P. M. | | | | | | | | | | | | | | | | | | |
| 1 | 2 | 5.8 | 0 | 0 | 2 | 5.8 | 0 | 0 | 0 | 0 | 0 | 0 | 0 | 0 | 0 | 0 | 0 | 0 |
| 2 | 3 | 8.7 | 2 | 7.0 | 5 | 15.7 | 0 | 0 | 0 | 0 | 0 | 0 | 0 | 0 | 0 | 0 | 0 | 0 |
| 3 | 2 | 4.1 | 2 | 5.0 | 4 | 9.1 | 0 | 0 | 3 | 8.0 | 3 | 8.0 | 1 | 5.0 | 0 | 0 | 1 | 5.0 |
| 4 | 2 | 9.1 | 1 | 2.0 | 3 | 11.1 | 1 | 3.0 | 4 | 13.0 | 5 | 16.0 | 3 | 11.2 | 1 | 2.5 | 4 | 13.7 |
| 5 | 2 | 4.3 | 1 | 2.8 | 3 | 7.1 | 0 | 0 | 3 | 11.1 | 3 | 11.1 | 3 | 7.9 | 1 | 2.3 | 4 | 10.2 |
| 6 | 2 | 8.8 | 0 | 0 | 2 | 8.8 | 4 | 16.5 | 4 | 11.5 | 8 | 28.0 | 4 | 16.3 | 2 | 4.4 | 6 | 20.7 |
| 7 | 1 | 16.0 | 1 | 2.0 | 2 | 18.0 | 2 | 7.0 | 0 | 0 | 2 | 7.0 | 2 | 15.0 | 2 | 4.9 | 4 | 19.9 |
| 8 | 2 | 20.3 | 0 | 0 | 2 | 20.3 | 3 | 9.3 | 1 | 2.0 | 4 | 11.3 | 5 | 26.3 | 2 | 5.0 | 7 | 31.3 |
| 9 | 1 | 6.0 | 0 | 0 | 1 | 6.0 | 1 | 2.0 | 0 | 0 | 1 | 2.0 | 1 | 3.3 | 2 | 7.3 | 3 | 10.6 |
| 10 | 4 | 14.7 | 1 | 2.6 | 5 | 17.3 | 4 | 12.5 | 1 | 3.0 | 5 | 15.0 | 4 | 12.4 | 0 | 0 | 4 | 12.4 |
| 11 | 3 | 14.0 | 0 | 0 | 3 | 14.0 | 0 | 0 | 0 | 0 | 0 | 0 | 2 | 12.5 | 0 | 0 | 2 | 12.5 |
| 12 | 3 | 9.0 | 1 | 2.0 | 4 | 11.0 | 0 | 0 | 1 | 6.0 | 1 | 6.0 | 4 | 12.7 | 2 | 5.2 | 6 | 17.9 |
| Total | 46 | 214.2 | 28 | 165.0 | 74 | 379.2 | 28 | 81.0 | 29 | 95.2 | 57 | 176.2 | 45 | 178.7 | 34 | 126.6 | 79 | 305.3 |

TABLE VIII.—*Disturbances in declination*—Continued.

In the columns headed No. are given the number of traces of declination for that month and hour which differ 2 minutes or more from the mean declination, as shown by the composite curve, for that month and hour. The columns headed Value give the sum of these differences in minutes of arc.

1888.

| Hour. | APRIL. | | | | | | MAY. | | | | | | JUNE. | | | | | |
|---|---|---|---|---|---|---|---|---|---|---|---|---|---|---|---|---|---|---|
| | Easterly. | | Westerly. | | Total. | | Easterly. | | Westerly. | | Total. | | Easterly. | | Westerly. | | Total. | |
| | No. | Value. | No. | Value. | No. | Value. | No. | Value. | No. | Value. | No. | Value. | No. | Value. | No | Value. | No. | Value. |
| A.M. | | | | | | | | | | | | | | | | | | |
| 1 | 5 | 16.0 | 2 | 12.0 | 7 | 28.0 | 5 | 26.3 | 1 | 2.3 | 6 | 28.6 | 5 | 15.0 | 0 | 0 | 5 | 15.0 |
| 2 | 2 | 7.0 | 2 | 4.3 | 4 | 11.3 | 1 | 13.3 | 3 | 10.8 | 4 | 24.1 | 5 | 18.4 | 3 | 13.0 | 8 | 31.4 |
| 3 | 2 | 9.9 | 4 | 10.2 | 6 | 20.1 | 5 | 24.9 | 4 | 12.0 | 9 | 36.9 | 2 | 7.0 | 2 | 9.0 | 4 | 16.0 |
| 4 | 4 | 20.5 | 4 | 10.5 | 8 | 31.0 | 4 | 17.4 | 4 | 17.5 | 8 | 34.9 | 2 | 6.8 | 2 | 5.5 | 4 | 15.3 |
| 5 | 1 | 7.0 | 4 | 15.6 | 5 | 22.6 | 1 | 3.4 | 3 | 11.0 | 4 | 14.4 | 4 | 14.8 | 2 | 7.0 | 6 | 21.8 |
| 6 | 2 | 6.5 | 2 | 11.5 | 4 | 18.0 | 1 | 3.4 | 0 | 0 | 1 | 0 | 2 | 9.8 | 1 | 4.2 | 3 | 14.0 |
| 7 | 0 | 0 | 1 | 4.0 | 1 | 4.0 | 2 | 6.4 | 1 | 5.7 | 3 | 12.1 | 2 | 5.5 | 0 | 0 | 2 | 5.5 |
| 8 | 0 | 0 | 3 | 18.5 | 3 | 18.5 | 4 | 8.5 | 1 | 10.0 | 5 | 18.5 | 2 | 7.2 | 1 | 3.0 | 3 | 10.2 |
| 9 | 0 | 0 | 3 | 12.8 | 3 | 12.8 | 1 | 4.0 | 1 | 10.0 | 2 | 14.0 | 0 | 0 | 1 | 4.0 | 1 | 4.0 |
| 10 | 0 | 0 | 3 | 10.6 | 3 | 10.6 | 1 | 5.2 | 0 | 0 | 1 | 5.2 | 1 | 4.3 | 3 | 19.0 | 4 | 23.3 |
| 11 | 0 | 0 | 0 | 0 | 0 | 0 | 0 | 0 | 0 | 0 | 0 | 0 | 0 | 0 | 0 | 0 | 0 | 0 |
| 12 | 0 | 0 | 0 | 0 | 0 | 0 | 0 | 0 | 0 | 0 | 0 | 0 | 0 | 0 | 0 | 0 | 0 | 0 |
| P.M. | | | | | | | | | | | | | | | | | | |
| 1 | 0 | 0 | 0 | 0 | 0 | 0 | 1 | 4.5 | 0 | 0 | 1 | 4.5 | 2 | 5.9 | 2 | 6.5 | 4 | 12.4 |
| 2 | 0 | 0 | 2 | 10.0 | 2 | 10.0 | 1 | 3.5 | 1 | 3.6 | 2 | 7.1 | 0 | 0 | 1 | 10.7 | 1 | 10.7 |
| 3 | 1 | 5.3 | 4 | 16.0 | 5 | 21.3 | 1 | 3.0 | 1 | 5.6 | 2 | 8.6 | 2 | 6.3 | 1 | 7.1 | 3 | 13.4 |
| 4 | 0 | 0 | 6 | 25.1 | 6 | 25.1 | 1 | 4.8 | 1 | 3.2 | 2 | 8.0 | 0 | 0 | 1 | 6.5 | 1 | 6.5 |
| 5 | 0 | 0 | 4 | 16.9 | 4 | 16.9 | 1 | 3.2 | 1 | 2.5 | 2 | 5.7 | 0 | 0 | 1 | 17.5 | 1 | 17.5 |
| 6 | 2 | 8.7 | 2 | 10.5 | 4 | 19.2 | 0 | 0 | 0 | 0 | 0 | 0 | 0 | 0 | 3 | 10.5 | 3 | 10.5 |
| 7 | 2 | 8.7 | 1 | 2.5 | 3 | 11.2 | 1 | 3.0 | 0 | 0 | 1 | 3.0 | 2 | 6.4 | 1 | 2.5 | 3 | 8.9 |
| 8 | 6 | 21.8 | 0 | 0 | 6 | 21.8 | 4 | 20.0 | 1 | 2.4 | 5 | 22.4 | 2 | 5.9 | 1 | 2.6 | 3 | 8.5 |
| 9 | 5 | 21.3 | 0 | 0 | 5 | 21.3 | 1 | 6.0 | 1 | 9.0 | 2 | 15.0 | 4 | 30.0 | 2 | 5.3 | 6 | 35.3 |
| 10 | 2 | 11.0 | 0 | 0 | 2 | 11.0 | 6 | 32.0 | 0 | 0 | 6 | 32.0 | 4 | 22.9 | 1 | 1.8 | 5 | 24.7 |
| 11 | 8 | 26.4 | 1 | 4.6 | 9 | 31.0 | 6 | 35.8 | 0 | 0 | 6 | 35.8 | 5 | 14.9 | 2 | 7.4 | 7 | 22.3 |
| 12 | 5 | 15.0 | 2 | 5.5 | 7 | 20.5 | 5 | 28.0 | 0 | 0 | 5 | 28.0 | 4 | 15.5 | 2 | 5.2 | 6 | 20.7 |
| Total. | 47 | 178.1 | 50 | 201.1 | 97 | 379.2 | 53 | 256.6 | 24 | 105.6 | 77 | 362.2 | 52 | 203.6 | 33 | 148.3 | 85 | 351.9 |

### TABLE VIII.—*Disturbances in declination*—Continued.

In the columns headed No. are given the number of traces of declination for that month and hour which differ 2 minutes or more from the mean declination, as shown by the composite curve, for that month and hour. The columns headed Value give the sum of these differences in minutes of arc.

1888.

| Hour. | July Easterly No. | Value. | Westerly No. | Value. | Total No. | Value. | August Easterly No. | Value. | Westerly No. | Value. | Total No. | Value | September Easterly No. | Value. | Westerly No. | Value. | Total No. | Value. |
|---|---|---|---|---|---|---|---|---|---|---|---|---|---|---|---|---|---|---|
| A.M. | | | | | | | | | | | | | | | | | | |
| 1 | 2 | 5.5 | 1 | 3.0 | 3 | 8.5 | 4 | 13.8 | 1 | 4.0 | 5 | 17.8 | 1 | 5.0 | 4 | 15.2 | 5 | 20.2 |
| 2 | 3 | 6.6 | 3 | 13.8 | 6 | 20.4 | 5 | 13.7 | 2 | 13.0 | 7 | 26.7 | 2 | 4.5 | 2 | 4.1 | 4 | 8.6 |
| 3 | 0 | 0 | 5 | 19.6 | 5 | 19.6 | 0 | 0 | 4 | 11.1 | 4 | 11.1 | 3 | 9.5 | 5 | 20.3 | 8 | 29.8 |
| 4 | 0 | 0 | 2 | 15.2 | 2 | 15.2 | 1 | 2.2 | 3 | 8.2 | 4 | 10.4 | 2 | 10.5 | 3 | 8.5 | 5 | 19.0 |
| 5 | 1 | 2.7 | 1 | 3.0 | 2 | 5.7 | 1 | 2.5 | 3 | 14.5 | 4 | 17.0 | 1 | 5.0 | 2 | 4.5 | 3 | 9.5 |
| 6 | 0 | 0 | 1 | 7.3 | 1 | 7.3 | 1 | 4.5 | 2 | 4.2 | 3 | 8.7 | 0 | 0 | 1 | 2.5 | 1 | 2.5 |
| 7 | 0 | 0 | 1 | 6.0 | 1 | 6.0 | 1 | 2.4 | 3 | 8.7 | 4 | 11.1 | 0 | 0 | 1 | 3.0 | 1 | 3.0 |
| 8 | 1 | 3.5 | 1 | 3.5 | 2 | 7.0 | 0 | 0 | 3 | 12.6 | 3 | 12.6 | 0 | 0 | 3 | 10.5 | 3 | 10.5 |
| 9 | 2 | 6.0 | 1 | 3.0 | 3 | 9.0 | 2 | 6.0 | 2 | 6.0 | 4 | 12.0 | 0 | 0 | 3 | 13.0 | 3 | 13.0 |
| 10 | 1 | 3.0 | 1 | 3.5 | 2 | 6.5 | 2 | 6.8 | 0 | 0 | 2 | 6.8 | 0 | 0 | 2 | 7.5 | 2 | 7.5 |
| 11 | 1 | 4.5 | 2 | 6.0 | 3 | 10.5 | 2 | 6.7 | 0 | 0 | 2 | 6.7 | 0 | 0 | 2 | 9.0 | 2 | 9.0 |
| 12 | 0 | 0 | 0 | 0 | 0 | 0 | 1 | 2.5 | 2 | 6.2 | 3 | 8.7 | 0 | 0 | 0 | 0 | 0 | 0 |
| P.M. | | | | | | | | | | | | | | | | | | |
| 1 | 0 | 0 | 2 | 5.6 | 2 | 5.6 | 1 | 2.5 | 2 | 6.7 | 3 | 9.2 | 0 | 0 | 1 | 3.0 | 1 | 3.0 |
| 2 | 0 | 0 | 2 | 7.4 | 2 | 7.4 | 0 | 0 | 1 | 3.0 | 1 | 3.0 | 0 | 0 | 0 | 0 | 0 | 0 |
| 3 | 0 | 0 | 2 | 6.0 | 2 | 6.0 | 2 | 5.2 | 1 | 3.0 | 3 | 8.2 | 1 | 3.0 | 0 | 0 | 1 | 3.0 |
| 4 | 0 | 0 | 1 | 3.8 | 1 | 3.8 | 0 | 0 | 2 | 5.2 | 2 | 5.2 | 1 | 2.5 | 0 | 0 | 1 | 2.5 |
| 5 | 1 | 3.0 | 1 | 4.0 | 2 | 7.0 | 0 | 0 | 1 | 2.5 | 1 | 2.5 | 1 | 4.0 | 1 | 4.0 | 2 | 8.0 |
| 6 | 1 | 5.0 | 1 | 3.0 | 2 | 8.0 | 3 | 10.7 | 2 | 4.2 | 5 | 14.9 | 2 | 11.0 | 1 | 2.8 | 3 | 13.8 |
| 7 | 1 | 3.0 | 0 | 0 | 1 | 3.0 | 4 | 13.7 | 2 | 5.2 | 6 | 18.9 | 3 | 8.0 | 1 | 2.0 | 4 | 10.0 |
| 8 | 2 | 9.8 | 0 | 0 | 2 | 9.8 | 6 | 24.2 | 0 | 0 | 6 | 30.2 | 5 | 32.0 | 0 | 0 | 5 | 32.0 |
| 9 | 6 | 23.3 | 1 | 1.5 | 7 | 24.8 | 4 | 21.0 | 0 | 0 | 4 | 21.0 | 5 | 26.5 | 0 | 0 | 5 | 26.5 |
| 10 | 5 | 16.2 | 0 | 0 | 5 | 16.2 | 1 | 3.5 | 0 | 0 | 1 | 3.5 | 3 | 13.0 | 0 | 0 | 3 | 13.0 |
| 11 | 1 | 5.5 | 0 | 0 | 1 | 5.5 | 4 | 32.5 | 0. | 0 | 4 | 32.5 | 0 | 0 | 0 | 0 | 0 | 0 |
| 12 | 3 | 4.9 | 1 | 3.5 | 4 | 8.4 | 3 | 13.5 | 1 | 2.5 | 4 | 16.0 | 2 | 7.5 | 2 | 6.2 | 4 | 13.7 |
| Total. | 31 | 102.5 | 30 | 118.7 | 61 | 221.2 | 48 | 187.9 | 37 | 120.7 | 85 | 308.6 | 32 | 142.0 | 34 | 116.1 | 66 | 258.1 |

### TABLE VIII.—*Disturbances in declination*—Continued.

In the columns headed No. are given the number of traces of declination for that month and hour which differ 2 minutes or more from the mean declination, as shown by the composite curve, for that month and hour. The columns headed Value give the sum of these differences in minutes of arc.

1888.

| Hour. | October. | | | | | | November. | | | | | | December. | | | | | |
|---|---|---|---|---|---|---|---|---|---|---|---|---|---|---|---|---|---|---|
| | Easterly. | | Westerly. | | Total. | | Easterly. | | Westerly. | | Total. | | Easterly. | | Westerly. | | Total. | |
| | No. | Value. | No. | Value. | No. | Value. | No. | Value. | No. | Value. | No. | Value. | No. | Value. | No. | Value. | No. | Value. |
| A. M. | | | | | | | | | | | | | | | | | | |
| 1 | 1 | 3.0 | 1 | 3.0 | 2 | 6.0 | 0 | 0 | 2 | 8.6 | 2 | 8.6 | 1 | 3.0 | 1 | 2.7 | 2 | 5.7 |
| 2 | 2 | 8.5 | 3 | 15.0 | 5 | 23.5 | 2 | 5.5 | 0 | 0 | 2 | 5.5 | 1 | 3.0 | 3 | 7.0 | 4 | 10.0 |
| 3 | 1 | 4.0 | 2 | 8.5 | 3 | 12.5 | 0 | 0 | 1 | 6.5 | 1 | 6.5 | 0 | 0 | 3 | 7.5 | 3 | 7.5 |
| 4 | 3 | 10.8 | 1 | 6.0 | 4 | 16.8 | 0 | 0 | 2 | 7.5 | 2 | 7.5 | 1 | 2.8 | 1 | 2.5 | 2 | 5.3 |
| 5 | 2 | 9.0 | 3 | 10.5 | 5 | 19.5 | 0 | 0 | 2 | 9.0 | 2 | 9.0 | 1 | 5.0 | 2 | 6.2 | 3 | 11.2 |
| 6 | 1 | 7.0 | 3 | 16.5 | 4 | 23.5 | 1 | 2.2 | 3 | 14.0 | 4 | 16.2 | 1 | 2.5 | 2 | 11.5 | 3 | 14.0 |
| 7 | 1 | 3.0 | 1 | 3.0 | 2 | 6.0 | 1 | 2.2 | 1 | 24.0 | 2 | 26.2 | 0 | 0 | 1 | 5.0 | 1 | 5.0 |
| 8 | 1 | 2.5 | 2 | 6.5 | 3 | 9.0 | 0 | 0 | 2 | 6.0 | 2 | 6.0 | 0 | 0 | 2 | 8.0 | 2 | 8.0 |
| 9 | 0 | 0 | 3 | 11.7 | 3 | 11.7 | 0 | 0 | 2 | 8.0 | 2 | 8.0 | 0 | 0 | 2 | 10.2 | 2 | 10.2 |
| 10 | 1 | 3.0 | 2 | 10.3 | 3 | 13.3 | 0 | 0 | 1 | 2.5 | 1 | 2.5 | 0 | 0 | 5 | 19.0 | 5 | 19.0 |
| 11 | 1 | 2.8 | 0 | 0 | 1 | 2.8 | 1 | 2.5 | 3 | 13.0 | 4 | 15.5 | 0 | 0 | 2 | 8.5 | 2 | 8.5 |
| 12 | 1 | 3.0 | 1 | 3.6 | 2 | 6.6 | 0 | 0 | 2 | 11.5 | 2 | 11.5 | 0 | 0 | 2 | 8.0 | 2 | 8.0 |
| P. M. | | | | | | | | | | | | | | | | | | |
| 1 | 1 | 3.0 | 1 | 4.8 | 2 | 7.8 | 1 | 2.5 | 1 | 2.5 | 2 | 5.0 | 0 | 0 | 1 | 2.1 | 1 | 2.0 |
| 2 | 0 | 0 | 3 | 12.0 | 3 | 12.0 | 0 | 0 | 1 | 2.5 | 1 | 2.5 | 0 | 0 | 1 | 3.0 | 1 | 3.0 |
| 3 | 0 | 0 | 3 | 8.8 | 3 | 8.8 | 0 | 0 | 1 | 2.6 | 1 | 2.6 | 0 | 0 | 0 | 0 | 0 | 0 |
| 4 | 0 | 0 | 3 | 10.5 | 3 | 10.5 | 1 | 4.0 | 0 | 0 | 1 | 4.0 | 0 | 0 | 1 | 3.0 | 1 | 3.0 |
| 5 | 1 | 4.0 | 2 | 9.5 | 3 | 13.5 | 0 | 0 | 0 | 0 | 0 | 0 | 0 | 0 | 0 | 0 | 0 | 0 |
| 6 | 3 | 20.5 | 2 | 8.5 | 5 | 29.0 | 0 | 0 | 0 | 0 | 0 | 0 | 0 | 0 | 0 | 0 | 0 | 0 |
| 7 | 3 | 11.0 | 0 | 0 | 3 | 11.0 | 2 | 9.0 | 1 | 3.0 | 3 | 12.0 | 0 | 0 | 0 | 0 | 0 | 0 |
| 8 | 4 | 11.0 | 0 | 0 | 4 | 11.0 | 2 | 13.0 | 0 | 0 | 2 | 13.0 | 1 | 2.5 | 0 | 0 | 1 | 2.5 |
| 9 | 2 | 18.5 | 0 | 0 | 2 | 18.5 | 2 | 4.5 | 0 | 0 | 2 | 4.5 | 3 | 12.5 | 0 | 0 | 3 | 12.5 |
| 10 | 3 | 12.5 | 0 | 0 | 3 | 12.5 | 0 | 0 | 1 | 2.0 | 1 | 2.0 | 2 | 7.0 | 0 | 0 | 2 | 7.0 |
| 11 | 4 | 25.4 | 1 | 4.0 | 5 | 29.4 | 0 | 0 | 1 | 9.0 | 1 | 9.0 | 3 | 33.0 | 0 | 0 | 3 | 33.0 |
| 12 | 2 | 8.2 | 0 | 0 | 2 | 8.2 | 3 | 9.7 | 2 | 3.5 | 5 | 13.2 | 2 | 5.5 | 1 | 2.0 | 4 | 7.5 |
| Total. | 38 | 173.7 | 37 | 152.7 | 75 | 326.4 | 16 | 55.1 | 29 | 135.7 | 45 | 190.8 | 16 | 76.8 | 30 | 106.1 | 46 | 182.9 |

## TABLE VIII.—*Disturbances in declination*—Continued.

In the columns headed No. are given the number of traces of declination for that month and hour which differ 2 minutes or more from the mean declination, as shown by the composite curve, for that month and hour. The columns headed Value give the sum of these differences in minutes of arc.

1889.

| Hour. | JANUARY. Easterly. No. | Value. | Westerly. No. | Value. | Total. No. | Value. | FEBRUARY. Easterly. No. | Value. | Westerly. No. | Value. | Total. No. | Value. | MARCH. Easterly. No. | Value. | Westerly. No. | Value. | Total. No. | Value. |
|---|---|---|---|---|---|---|---|---|---|---|---|---|---|---|---|---|---|---|
| A. M. | | | | | | | | | | | | | | | | | | |
| 1 | 1 | 2.9 | 0 | 0 | 1 | 2.9 | 2 | 5.5 | 1 | 5.0 | 3 | 10.5 | 2 | 10.0 | 1 | 2.1 | 3 | 12.1 |
| 2 | 1 | 7.5 | 0 | 0 | 1 | 7.5 | 1 | 2.8 | 2 | 4.0 | 3 | 6.8 | 4 | 12.8 | 3 | 6.5 | 7 | 19.3 |
| 3 | 1 | 4.3 | 1 | 2.8 | 2 | 7.1 | 1 | 2.4 | 0 | 0 | 1 | 2.4 | 3 | 9.7 | 1 | 2.7 | 4 | 12.4 |
| 4 | 2 | 4.0 | 0 | 0 | 2 | 4.0 | 0 | 0 | 1 | 4.2 | 1 | 4.2 | 1 | 3.2 | 2 | 4.9 | 3 | 8.1 |
| 5 | 1 | 2.8 | 1 | 3.8 | 2 | 6.6 | 0 | 0 | 1 | 7.0 | 1 | 7.0 | 1 | 4.0 | 2 | 9.6 | 3 | 13.6 |
| 6 | 0 | 0 | 1 | 3.1 | 1 | 3.1 | 2 | 5.8 | 1 | 4.0 | 3 | 9.8 | 0 | 0 | 1 | 5.0 | 1 | 5.0 |
| 7 | 0 | 0 | 1 | 3.7 | 1 | 3.7 | 2 | 5.5 | 1 | 4.0 | 3 | 9.5 | 1 | 2.0 | 1 | 3.7 | 2 | 5.7 |
| 8 | 0 | 0 | 0 | 0 | 0 | 0 | 0 | 0 | 3 | 18.0 | 3 | 18.0 | 0 | 0 | 3 | 9.3 | 3 | 9.3 |
| 9 | 0 | 0 | 1 | 6.7 | 1 | 6.7 | 2 | 5.5 | 2 | 8.9 | 4 | 14.4 | 1 | 2.0 | 1 | 5.7 | 2 | 7.7 |
| 10 | 2 | 4.3 | 3 | 7.7 | 5 | 12.0 | 2 | 4.3 | 2 | 6.3 | 4 | 10.6 | 0 | 0 | 1 | 8.5 | 1 | 8.5 |
| 11 | 1 | 2.1 | 3 | 9.1 | 4 | 11.2 | 1 | 2.0 | 3 | 10.3 | 4 | 12.3 | 0 | 0 | 2 | 7.1 | 2 | 7.1 |
| 12 | 0 | 0 | 3 | 8.1 | 3 | 8.1 | 0 | 0 | 3 | 7.9 | 3 | 7.9 | 2 | 4.1 | 2 | 8.2 | 4 | 12.3 |
| P. M. | | | | | | | | | | | | | | | | | | |
| 1 | 2 | 4.9 | 3 | 11.1 | 5 | 16.0 | 0 | 0 | 1 | 2.0 | 1 | 2.0 | 1 | 2.0 | 2 | 4.1 | 3 | 6.1 |
| 2 | 1 | 2.2 | 4 | 11.0 | 5 | 13.2 | 0 | 0 | 0 | 0 | 0 | 0 | 0 | 0 | 3 | 9.1 | 3 | 9.1 |
| 3 | 1 | 2.0 | 3 | 11.5 | 4 | 13.5 | 0 | 0 | 1 | 2.7 | 1 | 2.7 | 1 | 2.0 | 2 | 6.6 | 3 | 8.6 |
| 4 | 1 | 2.0 | 1 | 2.1 | 2 | 4.1 | 0 | 0 | 1 | 2.9 | 1 | 2.9 | 0 | 0 | 5 | 15.4 | 5 | 15.4 |
| 5 | 0 | 0 | 0 | 0 | 0 | 0 | 0 | 0 | 1 | 2.5 | 1 | 2.5 | 1 | 2.0 | 2 | 5.9 | 3 | 7.9 |
| 6 | 0 | 0 | 0 | 0 | 0 | 0 | 0 | 0 | 1 | 2.0 | 1 | 2.0 | 1 | 2.2 | 1 | 3.2 | 2 | 5.4 |
| 7 | 0 | 0 | 1 | 2.0 | 1 | 2.0 | 1 | 2.5 | 0 | 0 | 1 | 2.5 | 1 | 5.0 | 2 | 7.0 | 3 | 12.0 |
| 8 | 2 | 4.4 | 0 | 0 | 2 | 4.4 | 1 | 4.8 | 1 | 2.0 | 2 | 6.8 | 2 | 5.0 | 1 | 2.7 | 3 | 7.7 |
| 9 | 2 | 8.0 | 0 | 0 | 2 | 8.0 | 4 | 9.7 | 0 | 0 | 4 | 9.7 | 2 | 7.0 | 0 | 0 | 2 | 7.0 |
| 10 | 1 | 2.2 | 1 | 2.0 | 2 | 4.2 | 2 | 6.4 | 1 | 2.9 | 3 | 9.3 | 5 | 15.4 | 1 | 2.0 | 6 | 17.4 |
| 11 | 0 | 0 | 0 | 0 | 0 | 0 | 0 | 0 | 1 | 3.1 | 1 | 3.1 | 4 | 18.9 | 1 | 2.2 | 5 | 21.1 |
| 12 | 0 | 0 | 1 | 5.1 | 1 | 5.1 | 0 | 0 | 3 | 13.8 | 3 | 13.8 | 5 | 16.8 | 3 | 7.9 | 8 | 24.7 |
| Total . | 19 | 53.6 | 28 | 89.8 | 47 | 143.4 | 24 | 71.0 | 28 | 99.7 | 52 | 170.7 | 38 | 124.1 | 43 | 139.4 | 81 | 263.5 |

### TABLE VIII.—*Disturbances in declination*—Continued.

In the columns headed No. are given the number of traces of declination for that month and hour which differ 2 minutes or more from the mean declination, as shown by the composite curve, for that month and hour. The columns headed Value give the sum of these differences in minutes of arc.

1889.

| Hour. | APRIL Easterly No. | Value. | Westerly No. | Value. | Total No. | Value. | MAY Easterly No. | Value. | Westerly No. | Value. | Total No. | Value. | JUNE Easterly No. | Value. | Westerly No. | Value. | Total No. | Value. |
|---|---|---|---|---|---|---|---|---|---|---|---|---|---|---|---|---|---|---|
| A. M. | | | | | | | | | | | | | | | | | | |
| 1 | 1 | 3.0 | 3 | 7.0 | 4 | 10.0 | 5 | 18.6 | 3 | 11.5 | 8 | 30.1 | 4 | 11.3 | 2 | 5.3 | 6 | 16.6 |
| 2 | 2 | 6.8 | 2 | 5.0 | 4 | 11.8 | 5 | 19.2 | 5 | 19.0 | 10 | 38.2 | 0 | 0 | 2 | 5.5 | 2 | 5.5 |
| 3 | 2 | 14.2 | 2 | 6.0 | 4 | 20.2 | 3 | 12.2 | 4 | 11.4 | 7 | 23.6 | 0 | 0 | 1 | 2.2 | 1 | 2.2 |
| 4 | 0 | 0 | 2 | 7.0 | 2 | 7.0 | 4 | 11.4 | 4 | 14.4 | 8 | 25.8 | 4 | 14.0 | 3 | 10.1 | 7 | 24.1 |
| 5 | 0 | 0 | 2 | 4.2 | 2 | 4.2 | 4 | 15.7 | 2 | 7.4 | 6 | 23.1 | 4 | 10.0 | 3 | 14.4 | 7 | 24.4 |
| 6 | 1 | 3.1 | 1 | 2.1 | 2 | 5.2 | 4 | 15.9 | 4 | 11.8 | 8 | 27.7 | 4 | 11.5 | 3 | 8.0 | 7 | 19.5 |
| 7 | 2 | 5.3 | 1 | 2.1 | 3 | 7.4 | 1 | 4.8 | 4 | 12.6 | 5 | 17.4 | 2 | 8.6 | 2 | 10.1 | 4 | 18.7 |
| 8 | 3 | 9.6 | 3 | 9.4 | 6 | 19.0 | 1 | 3.0 | 3 | 9.4 | 4 | 12.4 | 3 | 9.9 | 4 | 20.8 | 7 | 30.7 |
| 9 | 3 | 7.3 | 3 | 8.1 | 6 | 15.4 | 2 | 5.3 | 4 | 12.2 | 6 | 17.5 | 4 | 15.0 | 3 | 12.9 | 7 | 27.9 |
| 10 | 2 | 6.9 | 2 | 6.8 | 4 | 13.7 | 3 | 8.0 | 4 | 12.9 | 7 | 20.9 | 3 | 11.0 | 3 | 16.4 | 6 | 27.4 |
| 11 | 5 | 16.4 | 2 | 5.4 | 7 | 21.8 | 2 | 5.2 | 4 | 15.9 | 6 | 21.1 | 1 | 3.6 | 3 | 11.5 | 4 | 15.1 |
| 12 | 3 | 9.9 | 3 | 7.5 | 6 | 17.4 | 1 | 3.0 | 2 | 11.5 | 3 | 14.5 | 1 | 4.0 | 2 | 12.5 | 3 | 16.5 |
| P. M. | | | | | | | | | | | | | | | | | | |
| 1 | 4 | 11.9 | 4 | 10.3 | 8 | 22.2 | 1 | 2.8 | 3 | 8.7 | 4 | 11.5 | 3 | 7.3 | 3 | 12.1 | 6 | 19.4 |
| 2 | 3 | 9.0 | 2 | 5.1 | 5 | 14.1 | 3 | 7.3 | 1 | 3.4 | 4 | 10.7 | 1 | 2.2 | 4 | 13.8 | 5 | 16.0 |
| 3 | 2 | 4.2 | 1 | 2.2 | 3 | 6.4 | 3 | 8.3 | 3 | 7.3 | 6 | 15.6 | 2 | 4.9 | 2 | 7.4 | 4 | 12.3 |
| 4 | 1 | 2.0 | 3 | 7.6 | 4 | 9.6 | 3 | 8.7 | 4 | 10.5 | 7 | 19.2 | 1 | 2.1 | 2 | 8.4 | 3 | 10.5 |
| 5 | 1 | 2.0 | 2 | 4.7 | 3 | 6.7 | 2 | 5.0 | 2 | 4.8 | 4 | 9.8 | 0 | 0 | 3 | 11.3 | 3 | 11.3 |
| 6 | 0 | 0 | 1 | 2.0 | 1 | 2.0 | 0 | 0 | 3 | 11.1 | 3 | 11.1 | 0 | 0 | 3 | 11.7 | 3 | 11.7 |
| 7 | 1 | 3.3 | 3 | 7.9 | 4 | 11.2 | 0 | 0 | 2 | 4.5 | 2 | 4.5 | 1 | 2.0 | 4 | 12.8 | 5 | 14.8 |
| 8 | 3 | 9.9 | 2 | 4.1 | 5 | 14.0 | 1 | 2.0 | 2 | 4.5 | 3 | 6.5 | 1 | 3.0 | 3 | 11.8 | 4 | 14.8 |
| 9 | 3 | 11.7 | 1 | 2.1 | 4 | 13.8 | 1 | 4.0 | 2 | 6.2 | 3 | 10.2 | 2 | 5.8 | 2 | 9.0 | 4 | 14.8 |
| 10 | 4 | 33.8 | 3 | 7.1 | 7 | 40.9 | 4 | 19.1 | 2 | 6.0 | 6 | 25.1 | 3 | 8.7 | 3 | 10.2 | 6 | 18.9 |
| 11 | 5 | 19.2 | 2 | 5.1 | 7 | 24.3 | 2 | 11.0 | 1 | 2.6 | 3 | 13.6 | 3 | 7.1 | 3 | 6.9 | 6 | 14.0 |
| 12 | 2 | 13.1 | 2 | 6.7 | 4 | 19.8 | 4 | 16.2 | 2 | 5.4 | 6 | 21.6 | 2 | 7.1 | 1 | 2.0 | 3 | 9.1 |
| Total | 53 | 202.6 | 52 | 135.5 | 105 | 338.1 | 59 | 206.7 | 70 | 225.0 | 129 | 431.7 | 49 | 149.1 | 64 | 247.1 | 113 | 396.2 |

## TABLE VIII.—*Disturbances in declination*—Continued.

In the columns headed No. are given the number of traces of declination for that month and hour which differ 2 minutes or more from the mean declination, as shown by the composite curve, for that month and hour. The columns headed Value give the sum of these differences in minutes of arc.

### 1889.

| Hour | July Easterly | | July Westerly | | July Total | | August Easterly | | August Westerly | | August Total | | September Easterly | | September Westerly | | September Total | |
|---|---|---|---|---|---|---|---|---|---|---|---|---|---|---|---|---|---|---|
| | No. | Value. | No. | Value. | No. | Value. | No. | Value. | No. | Value. | No. | Value. | No. | Value. | No. | Value. | No. | Value. |
| **A. M.** | | | | | | | | | | | | | | | | | | |
| 1 | 3 | 13.2 | 1 | 2.5 | 4 | 15.7 | 2 | 8.0 | 4 | 12.0 | 6 | 20.0 | 0 | 0 | 3 | 17.5 | 3 | 17.5 |
| 2 | 3 | 16.0 | 0 | 0 | 3 | 16.0 | 2 | 5.4 | 3 | 8.8 | 5 | 14.2 | 2 | 4.6 | 4 | 18.5 | 6 | 23.1 |
| 3 | 2 | 23.4 | 1 | 5.0 | 3 | 28.4 | 2 | 5.4 | 6 | 18.4 | 8 | 23.8 | 1 | 3.5 | 6 | 25.1 | 7 | 28.6 |
| 4 | 1 | 6.6 | 4 | 12.0 | 5 | 18.6 | 1 | 3.4 | 3 | 9.0 | 4 | 12.4 | 0 | 0 | 4 | 17.2 | 4 | 17.2 |
| 5 | 0 | 0 | 0 | 0 | 0 | 0 | 1 | 3.0 | 2 | 9.6 | 3 | 12.6 | 1 | 2.0 | 6 | 23.1 | 7 | 25.1 |
| 6 | 0 | 0 | 0 | 0 | 0 | 0 | 3 | 12.5 | 2 | 13.2 | 5 | 25.7 | 2 | 6.9 | 4 | 21.0 | 6 | 27.9 |
| 7 | 0 | 0 | 0 | 0 | 0 | 0 | 2 | 9.0 | 2 | 10.0 | 4 | 19.0 | 1 | 2.0 | 3 | 14.6 | 4 | 16.6 |
| 8 | 1 | 3.5 | 2 | 7.4 | 3 | 10.9 | 2 | 8.9 | 1 | 2.8 | 3 | 11.7 | 0 | 0 | 3 | 7.2 | 3 | 7.2 |
| 9 | 1 | 3.5 | 0 | 0 | 1 | 3.5 | 4 | 14.1 | 2 | 8.6 | 6 | 22.7 | 1 | 2.3 | 2 | 6.8 | 3 | 9.1 |
| 10 | 2 | 6.5 | 0 | 0 | 2 | 6.5 | 3 | 10.4 | 1 | 5.0 | 4 | 15.4 | 2 | 5.3 | 3 | 10.9 | 5 | 16.2 |
| 11 | 0 | 0 | 1 | 3.1 | 1 | 3.1 | 1 | 4.4 | 2 | 8.0 | 3 | 12.4 | 3 | 7.5 | 2 | 11.0 | 5 | 18.5 |
| 12 | 0 | 0 | 3 | 10.8 | 3 | 10.8 | 0 | 0 | 2 | 6.9 | 2 | 6.9 | 0 | 0 | 2 | 7.0 | 2 | 7.0 |
| **P. M.** | | | | | | | | | | | | | | | | | | |
| 1 | 1 | 3.2 | 1 | 3.2 | 2 | 6.4 | 3 | 10.0 | 3 | 8.1 | 6 | 18.1 | 2 | 7.7 | 1 | 3.7 | 3 | 11.4 |
| 2 | 2 | 6.5 | 0 | 0 | 2 | 6.5 | 2 | 4.6 | 2 | 7.0 | 4 | 11.6 | 0 | 0 | 1 | 3.0 | 1 | 3.0 |
| 3 | 1 | 3.8 | 0 | 0 | 1 | 3.8 | 1 | 2.8 | 1 | 4.5 | 2 | 7.3 | 1 | 2.9 | 2 | 3.9 | 3 | 6.8 |
| 4 | 0 | 0 | 0 | 0 | 0 | 0 | 0 | 0 | 2 | 7.1 | 2 | 7.1 | 2 | 5.9 | 1 | 2.5 | 3 | 8.4 |
| 5 | 0 | 0 | 2 | 6.0 | 2 | 6.0 | 0 | 0 | 1 | 3.1 | 1 | 3.1 | 2 | 5.4 | 1 | 2.2 | 3 | 7.6 |
| 6 | 0 | 0 | 1 | 2.8 | 1 | 2.8 | 2 | 7.4 | 1 | 2.6 | 3 | 10.0 | 1 | 2.6 | 0 | 0 | 1 | 2.6 |
| 7 | 1 | 2.7 | 0 | 0 | 1 | 2.7 | 0 | 0 | 0 | 0 | 0 | 0 | 3 | 34.0 | 0 | 0 | 3 | 34.0 |
| 8 | 2 | 12.5 | 0 | 0 | 2 | 12.5 | 0 | 0 | 0 | 0 | 0 | 0 | 6 | 28.0 | 2 | 4.3 | 8 | 32.3 |
| 9 | 4 | 20.8 | 0 | 0 | 4 | 20.8 | 4 | 14.0 | 0 | 0 | 4 | 14.0 | 1 | 3.5 | 0 | 0 | 1 | 3.5 |
| 10 | 4 | 14.6 | 0 | 0 | 4 | 14.6 | 3 | 9.5 | 0 | 0 | 3 | 9.5 | 2 | 5.5 | 0 | 0 | 2 | 5.5 |
| 11 | 5 | 16.2 | 1 | 4.2 | 6 | 20.4 | 4 | 12.7 | 1 | 2.0 | 5 | 14.7 | 2 | 4.8 | 1 | 2.7 | 3 | 7.5 |
| 12 | 5 | 14.3 | 0 | 0 | 5 | 14.3 | 3 | 10.6 | 0 | 0 | 3 | 10.6 | 0 | 0 | 1 | 7.6 | 1 | 7.6 |
| Total | 38 | 167.3 | 17 | 57.0 | 55 | 224.3 | 44 | 154.1 | 41 | 146.7 | 85 | 300.8 | 35 | 134.4 | 52 | 209.8 | 87 | 344.2 |

## TABLE VIII.—*Disturbances in declination*—Continued.

In the columns headed No. are given the number of traces of declination for that month and hour which differ 2 minutes or more from the mean declination, as shown by the composite curve, for that month and hour. The columns headed Value give the sum of these differences in minutes of arc.

1889.

| Hour. | OCTOBER. Easterly. | | Westerly. | | Total. | | NOVEMBER. Easterly. | | Westerly. | | Total. | | DECEMBER. Easterly. | | Westerly. | | Total. | |
|---|---|---|---|---|---|---|---|---|---|---|---|---|---|---|---|---|---|---|
| | No. | Value. | No. | Value. | No. | Value. | No. | Value. | No. | Value. | No. | Value. | No. | Value. | No. | Value. | No. | Value. |
| **A.M.** | | | | | | | | | | | | | | | | | | |
| 1 | 2 | 7.0 | 2 | 4.5 | 4 | 11.5 | 3 | 10.9 | 2 | 8.0 | 5 | 18.9 | 1 | 6.0 | 1 | 3.0 | 2 | 9.0 |
| 2 | 4 | 13.1 | 3 | 10.5 | 7 | 23.6 | 1 | 3.8 | 4 | 11.8 | 5 | 15.6 | 1 | 2.9 | 1 | 3.9 | 2 | 6.8 |
| 3 | 2 | 9.8 | 4 | 13.7 | 6 | 23.5 | 1 | 2.0 | 4 | 11.1 | 5 | 13.1 | 2 | 4.5 | 0 | 0 | 2 | 4.5 |
| 4 | 2 | 9.4 | 3 | 12.1 | 5 | 21.5 | 2 | 4.5 | 4 | 16.7 | 6 | 21.2 | 0 | 0 | 1 | 2.0 | 1 | 2.0 |
| 5 | 0 | 0 | 1 | 3.9 | 1 | 3.9 | 0 | 0 | 2 | 5.0 | 2 | 5.0 | 1 | 2.1 | 2 | 4.4 | 3 | 6.5 |
| 6 | 0 | 0 | 5 | 18.7 | 5 | 18.7 | 0 | 0 | 4 | 17.3 | 4 | 17.3 | 1 | 2.7 | 2 | 4.0 | 3 | 6.7 |
| 7 | 2 | 4.9 | 3 | 8.5 | 5 | 13.4 | 0 | 0 | 6 | 24.4 | 6 | 24.4 | 1 | 2.0 | 1 | 2.0 | 2 | 4.0 |
| 8 | 0 | 0 | 5 | 19.4 | 5 | 19.4 | 0 | 0 | 5 | 17.3 | 5 | 17.3 | 1 | 2.1 | 1 | 3.4 | 2 | 5.5 |
| 9 | 0 | 0 | 5 | 21.4 | 5 | 21.4 | 2 | 5.6 | 5 | 16.1 | 7 | 21.7 | 2 | 4.3 | 1 | 2.0 | 3 | 6.3 |
| 10 | 1 | 3.0 | 4 | 13.0 | 5 | 16.0 | 2 | 4.7 | 7 | 28.1 | 9 | 32.8 | 2 | 4.9 | 1 | 2.0 | 3 | 6.9 |
| 11 | 2 | 4.6 | 4 | 25.6 | 6 | 30.2 | 2 | 4.9 | 5 | 19.0 | 7 | 23.9 | 1 | 2.0 | 2 | 5.5 | 3 | 7.5 |
| 12 | 0 | 0 | 4 | 11.4 | 4 | 11.4 | 2 | 5.5 | 3 | 10.3 | 5 | 15.8 | 0 | 0 | 3 | 6.2 | 3 | 6.2 |
| **P.M.** | | | | | | | | | | | | | | | | | | |
| 1 | 2 | 4.1 | 2 | 4.6 | 4 | 8.7 | 0 | 0 | 2 | 5.0 | 2 | 5.0 | 1 | 2.0 | 1 | 2.2 | 2 | 4.2 |
| 2 | 3 | 6.9 | 1 | 2.2 | 4 | 9.1 | 1 | 2.0 | 2 | 6.0 | 3 | 8.0 | 0 | 0 | 1 | 3.2 | 1 | 3.2 |
| 3 | 1 | 3.9 | 1 | 2.9 | 2 | 6.8 | 1 | 2.2 | 3 | 12.3 | 4 | 14.5 | 0 | 0 | 1 | 2.2 | 1 | 2.2 |
| 4 | 0 | 0 | 2 | 8.0 | 2 | 8.0 | 0 | 0 | 4 | 16.1 | 4 | 16.1 | 0 | 0 | 1 | 2.0 | 1 | 2.0 |
| 5 | 1 | 2.2 | 4 | 14.0 | 5 | 16.2 | 0 | 0 | 2 | 9.0 | 2 | 9.0 | 0 | 0 | 0 | 0 | 0 | 0 |
| 6 | 3 | 9.4 | 1 | 2.0 | 4 | 11.4 | 2 | 5.6 | 2 | 12.4 | 4 | 18.0 | 1 | 2.3 | 0 | 0 | 1 | 2.3 |
| 7 | 1 | 2.6 | 0 | 0 | 1 | 2.6 | 1 | 15.0 | 3 | 14.0 | 4 | 29.0 | 0 | 0 | 0 | 0 | 0 | 0 |
| 8 | 4 | 26.1 | 0 | 0 | 4 | 26.1 | 3 | 15.0 | 1 | 2.0 | 4 | 17.0 | 1 | 10.7 | 0 | 0 | 1 | 10.7 |
| 9 | 3 | 14.9 | 0 | 0 | 3 | 14.9 | 6 | 20.8 | 1 | 2.4 | 7 | 23.2 | 2 | 6.7 | 0 | 0 | 2 | 6.7 |
| 10 | 3 | 8.2 | 2 | 6.8 | 5 | 15.0 | 4 | 12.1 | 1 | 2.6 | 5 | 14.7 | 4 | 12.4 | 0 | 0 | 4 | 12.4 |
| 11 | 2 | 9.3 | 1 | 3.3 | 3 | 12.6 | 3 | 7.9 | 2 | 8.5 | 5 | 16.4 | 2 | 6.5 | 0 | 0 | 2 | 6.5 |
| 12 | 2 | 4.2 | 0 | 0 | 2 | 4.2 | 1 | 3.0 | 5 | 17.2 | 6 | 20.2 | 3 | 9.9 | 1 | 2.1 | 4 | 12.0 |
| Total. | 40 | 143.6 | 57 | 206.5 | 97 | 350.1 | 37 | 125.5 | 79 | 292.6 | 116 | 418.1 | 27 | 84.0 | 21 | 50.1 | 48 | 134.1 |

### TABLE IX.—*Observations for horizontal intensity.*

| Date. | Time. | Time vib. | Temp. vib. C. | Deflection. .3 M. | Deflection. .4 M. | Temp. def. C. | H. for .3 M. | H. for .4 M. | Mean H. | Magnetometer (ELLIOTT BROS.). |
|---|---|---|---|---|---|---|---|---|---|---|
| **1888.** | h. m. | s. | ° | ° ′ ″ | ° ′ ″ | ° | | | | |
| Jan. 7 | 10 38 | 4. 2991 | 17. 6 | 16 26 37 | 6 52 27 | 17. 3 | .198751 | .198729 | .198740 | No. 42 |
| 14 | 12 15 | 4. 3034 | 17. 25 | 16 27 25 | 6 52 23 | 18. 4 | .198414 | .198244 | .198329 | 42 |
| 20 | 11 22 | 4. 2958 | 13. 3 | 16 29 15 | 6 53 02 | 14. 9 | .198568 | .198655 | .198612 | 42 |
| 26 | 11 33 | 4. 2966 | 11. 2 | 16 28 57 | 6 53 12 | 14. 95 | .198454 | .198472 | .198463 | 42 |
| Feb. 2 | 11 37 | 4. 2973 | 12. 05 | 16 28 37 | 6 53 05 | 15. 2 | .198481 | .198495 | .198488 | 42 |
| 8 | 12 10 | 4. 3014 | 17. 3 | 16 25 32 | 6 52 45 | 19. 3 | .198642 | .198436 | .198539 | 42 |
| 14 | 10 49 | 4. 2959 | 13. 45 | 16 26 42 | 6 52 15 | 16. 6 | .198733 | .198760 | .198746 | 42 |
| 21 | 11 41 | 4. 3023 | 17. 3 | 16 25 45 | 6 52 00 | 18. 65 | .198623 | .198614 | .198619 | 42 |
| 28 | 10 32 | 4. 2920 | 10. 35 | 16 31 25 | 6 53 35 | 10. 77 | .198591 | .198756 | .198674 | 42 |
| Mar. 6 | 11 28 | 4. 2973 | 15. 1 | 16 27 40 | 6 52 22 | 13. 45 | .198820 | .198907 | .198863 | 42 |
| 15 | 11 50 | 4. 3018 | 15. 6 | 16 28 18 | 6 53 13 | 15. 00 | .198495 | .198445 | .198470 | 42 |
| 20 | 1 22 | 4. 3081 | 21. 3 | 16 25 15 | 6 51 00 | 20. 0 | .198317 | .198504 | .198412 | 42 |
| 27 | 11 13 | 4. 3025 | 15. 53 | 16 27 28 | 6 52 52 | 16. 0 | .198467 | .198445 | .198467 | 42 |
| Apr. 3 | 11 51 | 4. 3039 | 17. 5 | 16 23 50 | 6 51 45 | 17. 36 | .198820 | .198685 | .198757 | 42 |
| 10 | 11 36 | 4. 3008 | 17. 36 | 16 24 00 | 6 51 15 | 20. 31 | .198783 | .198783 | .198783 | 42 |
| 17 | 11 28 | 4. 2996 | 14. 5 | 16 26 08 | 6 52 03 | 14. 75 | .198769 | .198789 | .198779 | 42 |
| 24 | 11 17 | 4. 3001 | 13. 9 | 16 26 28 | 6 52 07 | 14. 54 | .198699 | .198735 | .198717 | 42 |
| May 1 | 1 00 | 4. 3099 | 21. 1 | 16 24 35 | 6 51 43 | 20. 62 | .198477 | .198425 | .198451 | 42 |
| 8 | 10 23 | 4. 3086 | 19. 33 | 16 25 55 | 6 52 02 | 19. 45 | .198381 | .198386 | .198384 | 42 |
| 15 | 10 40 | 4. 2994 | 15. 2 | 16 24 50 | 6 52 02 | 16. 39 | .198863 | .198759 | .198811 | 42 |
| 22 | 10 54 | 4. 3021 | 16. 0 | 16 26 03 | 6 52 08 | 16. 89 | .198632 | .198626 | .198629 | 42 |
| 29 | 11 25 | 4. 3140 | 27. 25 | 16 19 28 | 6 49 10 | 27. 7 | .198746 | .198731 | .198738 | 42 |
| June 12 | 11 02 | 4. 3074 | 20. 25 | 16 23 25 | 6 51 02 | 25. 62 | .198388 | .198382 | .198385 | 42 |
| 19 | 10 58 | 4. 3140 | 27. 1 | 16 18 28 | 6 49 07 | 27. 13 | .198855 | .198825 | .198840 | 42 |
| 26 | 10 44 | 4. 3139 | 27. 8 | 16 19 28 | 6 49 27 | 28. 15 | .198736 | .198726 | .198732 | 42 |
| July 3 | 10 39 | 4. 3092 | 21. 5 | 16 20 52 | 6 50 10 | 22. 69 | .198774 | .198729 | .198751 | 42 |
| 10 | 10 55 | 4. 3117 | 22. 9 | 16 22 30 | 6 50 25 | 23. 0 | .198563 | .198619 | .198591 | 42 |
| 17 | 10 53 | 4. 3118 | 23. 2 | 16 21 53 | 6 50 28 | 23. 75 | .198564 | .198548 | .198556 | 42 |
| 24 | 11 38 | 4. 3207 | 27. 9 | 16 19 10 | 6 49 18 | 28. 16 | .198443 | .198461 | .198452 | 42 |
| 31 | 10 17 | 4. 3150 | 23. 4 | 16 21 20 | 6 50 05 | 24. 0 | .198502 | .198520 | .198511 | 42 |
| Aug. 7 | 10 46 | 4. 3205 | 29. 5 | 16 16 47 | 6 48 20 | 31. 05 | .198642 | .198630 | .198636 | 42 |
| 14 | 10 58 | 4. 3084 | 22. 0 | 16 21 25 | 6 50 08 | 22. 92 | .198771 | .198782 | .198776 | 42 |
| 21 | 10 02 | 4. 3181 | 25. 5 | 16 20 50 | 6 50 15 | 26. 2 | .198401 | .198329 | .198365 | 42 |
| 28 | 9 45 | 4. 3089 | 21. 3 | 16 22 38 | 6 50 55 | 21. 5 | .198673 | .198607 | .198640 | 42 |
| Sept. 4 | 11 41 | 4. 3107 | 20. 5 | 16 23 15 | 6 49 58 | 20. 8 | .198520 | .198750 | .198635 | 42 |
| 11 | 10 12 | 4. 3104 | 19. 6 | 16 23 43 | 6 51 08 | 19. 97 | .198492 | .198492 | .198492 | 42 |
| 18 | 10 43 | 4. 3111 | 22. 7 | 16 22 05 | 6 50 25 | 22. 38 | .198652 | .198660 | .198656 | 42 |
| 25 | 9 56 | 4. 3062 | 16. 4 | 16 25 20 | 6 51 42 | 16. 7 | .198527 | .198542 | .198534 | 42 |
| Oct. 2 | 11 00 | 4. 3067 | 17. 7 | 16 23 55 | 6 51 35 | 17. 96 | .198660 | .198565 | .198613 | 42 |
| 9 | 11 15 | 4. 3017 | 13. 9 | 16 26 35 | 6 52 05 | 15. 34 | .198550 | .198609 | .198579 | 42 |
| 16 | 11 27 | 4. 3060 | 19. 5 | 16 22 10 | 6 50 47 | 19. 56 | .198827 | .198763 | .198795 | 42 |
| 23 | 10 31 | 4. 3090 | 20. 2 | 16 22 35 | 6 51 10 | 20. 63 | .198637 | .198518 | .198577 | 42 |
| 30 | 10 11 | 4. 3029 | 16. 25 | 16 22 20 | 6 50 47 | 20. 7 | .198778 | .198719 | .198748 | 42 |

TABLE IX.— *Observations for horizontal intensity*—Continued.

| Date. | Time. | Time vib. | Temp. vib. C. | Deflection. .3 M. | .4 M. | Temp. def. C. | H. for .3 M. | H. for .4 M. | Mean H. | Magnetometer (ELLIOTT BROS.). |
|---|---|---|---|---|---|---|---|---|---|---|
| 1888. | h. m. | s. | ° | ° ′ ″ | ° ′ ″ | ° | | | | |
| Nov. 6 | 11 15 | 4.3161 | 24.4 | 16 20 35 | 6 49 35 | 25.33 | .198486 | .198545 | .198516 | No. 42 |
| 13 | 9 57 | 4.2987 | 14.83 | 16 24 50 | 6 51 22 | 17.5 | .198779 | .198829 | .198804 | 42 |
| 20 | 10 48 | 4.2998 | 15.4 | 16 27 30 | 6 52 48 | 15.15 | .198628 | .198575 | .198602 | 42 |
| 27 | 11 02 | 4.3049 | 17.43 | 16 23 07 | 6 51 27 | 18.07 | .198763 | .198626 | .198695 | 42 |
| Dec. 4 | 9 46 | 4.3000 | 15.47 | 16 24 45 | 6 51 28 | 17.12 | .198775 | .198794 | .198784 | 42 |
| 11 | 10 48 | 4.3059 | 19.5 | 16 22 45 | 6 51 00 | 19.7 | .198811 | .198755 | .198783 | 42 |
| 18 | 11 14 | 4.3098 | 19.38 | 16 24 13 | 6 50 55 | 19.63 | .198477 | .198559 | .198518 | 42 |
| 26 | 11 13 | 4.3080 | 20.1 | 16 23 40 | 6 50 52 | 20.45 | .198200 | .198248 | .198224 | 42 |
| 31 | 10 20 | 4.3059 | 21.3 | 16 22 00 | 6 50 25 | 20.93 | .198857 | .198857 | .198857 | 42 |
| 1889. | | | | | | | | | | |
| Jan. 8 | 11 46 | 3.92050 | 18.07 | 20 44 00 | 8 33 15 | 16.95 | .198198 | .198288 | .198243 | 69 |
| 15 | 11 03 | 3.92143 | 21.80 | 20 39 12 | 8 31 40 | 21.98 | .198720 | .198745 | .198732 | 69 |
| 22 | 10 48 | 3.92080 | 14.07 | 20 42 37.5 | 8 33 07.5 | 19.18 | .198280 | .198290 | .198285 | 69 |
| 29 | 11 20 | 3.92070 | 19.52 | 20 39 40 | 8 31 42.5 | 18.50 | .198793 | .198847 | .198820 | 69 |
| Feb. 5 | 11 38 | 3.92480 | 20.00 | 20 37 32.5 | 8 31 20 | 20.60 | .198659 | .198615 | .198637 | 69 |
| 12 | 11 00 | 3.92330 | 15.80 | 20 39 20 | 8 31 20 | 16.20 | .198611 | .198715 | .198663 | 69 |
| 19 | 11 04 | 3.92710 | 20.80 | 20 34 24 | 8 29 37.5 | 19.37 | .199117 | .198945 | .199031 | 69 |
| 26 | 11 09 | 3.92640 | 14.13 | 20 35 22.5 | 8 30 15 | 14.87 | .198757 | .198533 | .198645 | 69 |
| Mar. 6 | 11 14 | 3.92620 | 15.65 | 20 38 12.5 | 8 31 12.5 | 14.70 | .198635 | .198625 | .198630 | 69 |
| 13 | 11 58 | 3.93333 | 22.00 | 20 33 22.5 | 8 29 39.5 | 21.85 | .198576 | .198536 | .198556 | 69 |
| 19 | 10 48 | 3.92109 | 15.30 | 20 34 47.5 | 8 30 05 | 16.19 | .199058 | .199049 | .199053 | 69 |
| 26 | 11 07 | 4.30620 | 17.50 | 16 24 27.5 | 6 51 15 | 17.49 | .198569 | .198631 | .198600 | 42 |
| Apr. 2 | 4 08 | 4.30875 | 19.60 | 16 22 37.5 | 6 50 22.5 | 19.50 | .198639 | .198732 | .198685 | 42 |
| 9 | 3 10 | 4.30412 | 17.10 | 16 23 25 | 6 50 45 | 16.10 | .198780 | .198842 | .198811 | 42 |
| 16 | 2 27 | 4.30393 | 16.75 | 16 23 17.5 | 6 50 55 | 16.25 | .198815 | .198844 | .198827 | 42 |
| 23 | 2 37 | 4.30865 | 21.13 | 16 21 00 | 6 49 57.5 | 20.97 | .198808 | .198853 | .198830 | 42 |
| 30 | 11 17 | 4.30602 | 16.70 | 16 23 35 | 6 51 20 | 17.31 | .198652 | .198607 | .198630 | 42 |
| May 7 | 3 31 | 4.31653 | 26.93 | 16 17 55 | 6 48 35 | 26.55 | .198601 | .198769 | .198685 | 42 |
| 14 | 2 41 | 4.31343 | 25.80 | 16 17 42.5 | 6 48 37.5 | 25.25 | .198876 | .198947 | .198912 | 42 |
| 21 | 3 45 | 4.31064 | 21.70 | 16 22 00 | 6 50 20 | 21.29 | .198554 | .198627 | .198590 | 42 |
| 28 | 2 51 | 4.31130 | 25.20 | 16 18 15 | 6 48 40 | 24.80 | .198884 | .198990 | .198937 | 42 |
| June 5 | 10 26 | 4.31250 | 22.40 | 16 22 20 | 6 50 27.5 | 22.84 | .198418 | .198494 | .198456 | 42 |
| 11 | 3 49 | 4.31537 | 26.95 | 16 16 35 | 6 48 25 | 27.83 | .198788 | .198791 | .198789 | 42 |
| 26 | 9 51 | 4.315046 | 23.10 | 16 22 35 | 6 50 40 | 22.31 | .198324 | .198375 | .198350 | 42 |
| July 2 | 3 23 | 4.31736 | 26.10 | 16 19 15 | 6 49 20 | 26.56 | .198500 | .198520 | .198510 | 42 |
| 9 | 2 26 | 4.31900 | 30.80 | 16 15 00 | 6 48 20 | 30.73 | .199050 | .198915 | .198982 | 42 |
| 16 | 2 46 | 4.31523 | 26.00 | 16 23 05 | 6 49 05 | 26.06 | .198218 | .198700 | .198459 | 42 |
| 23 | 2 58 | 4.32074 | 30.00 | 16 16 00 | 6 48 15 | 29.99 | .198586 | .198571 | .198578 | 42 |
| 30 | 2 56 | 4.31764 | 27.50 | 16 16 40 | 6 48 15 | 26.82 | .198761 | .198812 | .198786 | 42 |
| Aug. 6 | 2 39 | 4.31301 | 25.30 | 16 17 35 | 6 49 15 | 25.00 | .198810 | .198710 | .198760 | 42 |
| 13 | 2 50 | 4.31477 | 25.30 | 16 16 45 | 6 48 30 | 25.96 | .198891 | .198890 | .198890 | 42 |
| 21 | 1 47 | 4.31912 | 28.90 | 16 15 50 | 6 47 15 | 29.53 | .198984 | .198908 | .198946 | 42 |
| 27 | 2 15 | 4.30865 | 21.10 | 16 19 40 | 6 49 40 | 21.67 | .198790 | .198795 | .198793 | 42 |

TABLE IX.—*Observations for horizontal intensity*—Continued.

| Date. | Time. | Time vib. | Temp. vib. C. | Deflection. | | | | Temp. def. C. | H. for .3 M. | H. for .4 M. | Mean H. | Magneto-meter (ELLIOTT BROS.). |
|---|---|---|---|---|---|---|---|---|---|---|---|---|
| | | | | .3 M. | | .4 M. | | | | | | |
| 1889. | h. m. | s. | ° | ° | ′ | ″ | ° ′ ″ | ° | | | | |
| Sept. 3 | 1 46 | 4.31329 | 23.70 | 16 18 45 | | 6 49 45 | | 24.00 | .198701 | .198595 | .198648 | No. 42 |
| 10 | 2 22 | 4.31755 | 23.85 | 16 20 15 | | 6 49 40 | | 23.36 | .198596 | .198429 | .198513 | 42 |
| 19 | 2 11 | 4.30394 | 16.50 | 16 22 12 | | 6 50 25 | | 17.05 | .198825 | .198873 | .198849 | 42 |
| 24 | 2 10 | 4.30963 | 18.50 | 16 22 30 | | 6 50 55 | | 18.76 | .198609 | .198559 | .198584 | 42 |
| Oct. 1 | 2 04 | 4.31755 | 25.50 | 16 17 35 | | 6 48 40 | | 26.53 | .198661 | .198668 | .198665 | 42 |
| 22 | 1 45 | 4.31088 | 21.00 | 16 20 45 | | 6 50 45 | | 19.91 | .198769 | .198587 | .198678 | 42 |
| 29 | 1 36 | 4.30656 | 17.00 | 16 21 50 | | 6 50 35 | | 16.97 | .198809 | .198773 | .198791 | 42 |
| Nov. 19 | 1 55 | 4.31014 | 19.25 | 16 22 15 | | 6 50 25 | | 19.48 | .198581 | .198626 | .198604 | 42 |
| 26 | 2 28 | 4.31171 | 17.90 | 16 25 30 | | 6 51 50 | | 17.41 | .198229 | .198253 | .198241 | 42 |
| Dec. 3 | 2 06 | 4.30393 | 15.30 | 16 23 30 | | 6 51 10 | | 16.69 | .198575 | .198585 | .198580 | 42 |
| 17 | 2 17 | 4.29851 | 10.50 | 16 26 35 | | 6 52 15 | | 10.95 | .198672 | .198725 | .198698 | 42 |

## TABLE X.

| Date. | Hour. | ° | ′ | ″ | Needle. | Date. | Hour. | ° | ′ | ″ | Needle. |
|---|---|---|---|---|---|---|---|---|---|---|---|
| 1888. | | | | | | 1888. | | | | | |
| Jan. 6 | 2.25 p. m. | 71 | 10 | 13 | 2 | May 11 | 12.10 p. m. | 71 | 07 | 05 | 6 |
| 14 | 1.45 p. m. | 71 | 11 | 22 | 2 | 14 | 2.11 p. m. | 71 | 06 | 17 | 2 |
| 17 | 2.36 p. m. | 71 | 09 | 32 | 2 | 14 | 11.55 a. m. | 71 | 06 | 11 | 2 |
| 17 | 3.18 p. m. | 71 | 06 | 41 | 1 | 16 | 11.15 a. m. | 71 | 08 | 11 | 5 |
| 17 | 1.52 p. m. | 71 | 04 | 28 | 1 | 18 | 11.55 a. m. | 71 | 08 | 05 | 6 |
| 19 | . . . . . | 71 | 09 | 17 | 2 | 21 | 11.50 a. m. | 71 | 07 | 46 | 2 |
| 26 | 2.55 p. m. | 71 | 10 | 52 | 2 | 23 | . . . . . | 71 | 08 | 00 | 5 |
| Feb. 1 | 11.30 a. m. | 71 | 09 | 49 | 2 | 25 | 11.51 a. m. | 71 | 07 | 24 | 6 |
| 7 | 12.33 p. m. | 71 | 08 | 24 | 2 | 28 | 11.45 a. m. | 71 | 08 | 19 | 2 |
| 14 | 12.37 p. m. | 71 | 07 | 45 | 2 | 30 | 9.30 a. m. | 71 | 08 | 02 | 5 |
| 21 | 12.40 p. m. | 71 | 09 | 07 | 2 | June 1 | 11.30 a. m. | 71 | 06 | 17 | 6 |
| 25 | 1.00 p. m. | 71 | 10 | 19 | 2 | 4 | 12.00 m. | 71 | 08 | 50 | 2 |
| Mar. 6 | 3.22 p. m. | 71 | 10 | 07 | 2 | 6 | 12.05 p. m. | 71 | 07 | 58 | 5 |
| 9 | 2.10 p. m. | 71 | 08 | 45 | 2 | 11 | 12.00 m. | 71 | 06 | 22 | 2 |
| 9 | 11.32 a. m. | 71 | 08 | 52 | 5 | 13 | 12.02 p. m. | 71 | 07 | 14 | 5 |
| 9 | . . . . . | 71 | 08 | 47 | 6 | 15 | 12.52 p. m. | 71 | 06 | 58 | 6 |
| 12 | 12.32 p. m. | 71 | 10 | 09 | 2 | 18 | 12.15 p. m. | 71 | 06 | 34 | 2 |
| 14 | 12.28 p. m. | 71 | 08 | 29 | 5 | 20 | 11.50 a. m. | 71 | 06 | 13 | 5 |
| 16 | 12.57 p. m. | 71 | 10 | 40 | 6 | 22 | 12.10 p. m. | 71 | 07 | 17 | 6 |
| 19 | 12.20 p. m. | 71 | 08 | 49 | 2 | 25 | 12.00 m. | 71 | 06 | 38 | 2 |
| 21 | 2.15 p. m. | 71 | 10 | 11 | 5 | 27 | 11.57 a. m. | 71 | 06 | 10 | 5 |
| 23 | . . . . . | 71 | 08 | 02 | 6 | 29 | 11.55 a. m. | 71 | 06 | 24 | 6 |
| 26 | 12.20 p. m. | 71 | 08 | 24 | 2 | July 2 | 1.50 p. m. | 71 | 08 | 23 | 2 |
| 28 | 11.34 a. m. | 71 | 08 | 56 | 5 | 4 | 12.00 m. | 71 | 08 | 46 | 5 |
| 30 | 11.50 a. m. | 71 | 08 | 17 | 6 | 6 | 11.50 a. m. | 71 | 06 | 16 | 6 |
| Apr. 2 | 12.46 p. m. | 71 | 08 | 05 | 2 | 9 | 12.10 p. m. | 71 | 05 | 52 | 2 |
| 4 | 12.18 p. m. | 71 | 08 | 28 | 5 | 11 | 11.40 a. m. | 71 | 06 | 49 | 5 |
| 6 | 12.10 p. m. | 71 | 09 | 04 | 6 | 13 | 12.48 p. m. | 71 | 09 | 58 | 6 |
| 9 | 12.01 p. m. | 71 | 08 | 34 | 2 | 16 | 11.42 a. m. | 71 | 05 | 58 | 2 |
| 11 | 2.15 p. m. | 71 | 10 | 30 | 5 | 18 | 12.10 p. m. | 71 | 07 | 07 | 5 |
| 11 | . . . . . | 71 | 10 | 43 | 5 | 20 | 12.50 p. m. | 71 | 05 | 44 | 6 |
| 11 | 2.55 p. m. | 71 | 09 | 13 | 2 | 23 | 11.52 a. m. | 71 | 07 | 41 | 2 |
| 13 | 11.23 a. m. | 71 | 10 | 09 | 5 | 25 | 11.50 a. m. | 71 | 05 | 56 | 5 |
| 13 | 12.02 p. m. | 71 | 10 | 52 | 6 | 27 | 11.45 a. m. | 71 | 06 | 00 | 6 |
| 16 | 12.05 p. m. | 71 | 09 | 00 | 2 | Aug. 1 | 11.52 a. m. | 71 | 09 | 07 | 5 |
| 18 | 2.37 p. m. | 71 | 09 | 01 | 5 | 3 | . . . . . | 71 | 07 | 50 | 6 |
| 20 | 12.15 p. m. | 71 | 08 | 45 | 6 | 6 | 12.34 p. m. | 71 | 07 | 13 | 5 |
| 23 | 12.21 p. m. | 71 | 08 | 35 | 2 | 9 | 11.49 a. m. | 71 | 08 | 02 | 6 |
| 25 | 11.55 a. m. | 71 | 08 | 17 | 5 | 13 | 11.53 a. m. | 71 | 09 | 00 | 5 |
| 27 | 11.56 a. m. | 71 | 09 | 07 | 6 | 16 | 11.52 a. m. | 71 | 10 | 54 | 6 |
| 30 | 12.02 p. m. | 71 | 11 | 01 | 2 | 20 | 11.52 a. m. | 71 | 08 | 01 | 5 |
| May 2 | 12.10 p. m. | 71 | 07 | 26 | 5 | 23 | . . . . . | 71 | 07 | 49 | 6 |
| 4 | 11.45 a. m. | 71 | 07 | 32 | 6 | 27 | 12.01 p. m. | 71 | 08 | 02 | 5 |
| 7 | 12.33 p. m. | 71 | 08 | 20 | 2 | 30 | 11.50 a. m. | 71 | 07 | 10 | 6 |
| 9 | 11.59 a. m. | 71 | 08 | 49 | 5 | Sept. 3 | 12.23 p. m. | 71 | 06 | 45 | 5 |

9230——7

| 1888. | | ° | ′ | ″ | | 1889. | | ° | ′ | ″ | |
|---|---|---|---|---|---|---|---|---|---|---|---|
| Sept. 6 | 11.40 a. m. | 71 | 08 | 28 | 6 | Jan. 11 | 3.57 p. m. | 71 | 05 | 54 | 5ª |
| 10 | 11.40 a. m. | 71 | 06 | 54 | 5 | 12 | 1.30 p. m. | 71 | 06 | 31 | 5 |
| 10 | . . . . . | 71 | 07 | 43 | 6 | 12 | 2.15 p. m. | 71 | 06 | 39 | 6ª |
| 13 | 11.19 a. m. | 71 | 10 | 39 | 6 | 12 | 1.50 p. m. | 71 | 07 | 13 | 5ª |
| 13 | 12.12 p. m. | 71 | 09 | 39 | 5 | 14 | 11.34 a. m. | 71 | 07 | 39 | 5 |
| 17 | 11.38 a. m. | 71 | 08 | 45 | 5 | 16 | 11.54 a. m. | 71 | 07 | 34 | 5ª |
| 20 | 12.05 p. m. | 71 | 06 | 20 | 6 | 18 | . . . . . | 71 | 08 | 13 | 6ª |
| 24 | 11.23 a. m. | 71 | 08 | 35 | 5 | 22 | 11.57 a. m. | 71 | 08 | 02 | 5 |
| 24 | 11.51 a. m. | 71 | 06 | 58 | 6 | 24 | 11.57 a. m. | 71 | 07 | 07 | 5ª |
| 27 | 11.34 a. m. | 71 | 05 | 58 | 6 | 25 | 11.36 a. m. | 71 | 07 | 32 | 6ª |
| 27 | 12.17 p. m. | 71 | 07 | 00 | 5 | 28 | 12.02 p. m. | 71 | 08 | 00 | 5 |
| Oct. 1 | 11.57 a. m. | 71 | 07 | 09 | 5 | 30 | . . . . . | 71 | 06 | 56 | 5ª |
| 4 | 11.45 a. m. | 71 | 04 | 34 | 6 | Feb. 1 | 11.45 a. m. | 71 | 07 | 39 | 6ª |
| 8 | 3.11 p. m. | 71 | 06 | 52 | 5 | 4 | 10.35 a. m. | 71 | 06 | 02 | 5 |
| 11 | . . . . . | 71 | 06 | 41 | 5 | 6 | 11.43 a. m. | 71 | 06 | 24 | 5ª |
| 15 | . . . . . | 71 | 08 | 26 | 5 | 8 | 11.29 a. m. | 71 | 07 | 26 | 6ª |
| 18 | 11.44 a. m. | 71 | 05 | 58 | 5 | 11 | 11.32 a. m. | 71 | 08 | 22 | 5 |
| 20 | 12.12 p. m. | 71 | 08 | 13 | 5 | 13 | 11.26 a. m. | 71 | 06 | 13 | 5ª |
| 20 | 11.48 a. m. | 71 | 08 | 34 | 1 | 15 | 11.31 a. m. | 71 | 07 | 50 | 6ª |
| 20 | 11.25 a. m. | 71 | 07 | 04 | 2 | 18 | 11.32 a. m. | 71 | 07 | 30 | 5 |
| 22 | . . . . . | 71 | 07 | 56 | 5 | 20 | 11.35 a. m. | 71 | 08 | 43 | 5ª |
| 25 | 11.38 a. m. | 71 | 08 | 00 | 5 | 22 | 11.32 a. m. | 71 | 08 | 01 | 6ª |
| 29 | 12.27 p. m. | 71 | 08 | 07 | 5 | 25 | 11.35 a. m. | 71 | 06 | 09 | 5 |
| Nov. 1 | . . . . . | 71 | 06 | 37 | 5 | 27 | . . . . . | 71 | 06 | 00 | 5ª |
| 5 | . . . . . | 71 | 06 | 28 | 5 | Mar. 1 | 11.35 a. m. | 71 | 08 | 00 | 6ª |
| 8 | . . . . . | 71 | 06 | 43 | 5 | 5 | 11.38 a. m. | 71 | 07 | 49 | 5 |
| 12 | . . . . . | 71 | 08 | 09 | 5 | 7 | . . . . . | 71 | 06 | 05 | 5ª |
| 15 | 12.39 p. m. | 71 | 07 | 49 | 5 | 9 | . . . . . | 71 | 07 | 39 | 6ª |
| 20 | 12.05 p. m. | 71 | 09 | 22 | 5 | 11 | . . . . . | 71 | 08 | 02 | 5 |
| 22 | 11.40 a. m. | 71 | 07 | 09 | 5 | 15 | 11.39 a. m. | 71 | 07 | 47 | 6ª |
| 26 | 11.56 a. m. | 71 | 07 | 58 | 5 | 18 | 11.39 a. m. | 71 | 08 | 26 | 5 |
| Dec. 1 | 2.50 p. m. | 71 | 08 | 31 | 5 | 20 | 11.32 a. m. | 71 | 07 | 17 | 5ª |
| 3 | . . . . . | 71 | 07 | 22 | 5 | 22 | 11.40 a. m. | 71 | 09 | 00 | 6ª |
| 6 | . . . . . | 71 | 08 | 22 | 5 | 25 | 11.36 a. m. | 71 | 06 | 11 | 5 |
| 10 | . . . . . | 71 | 08 | 15 | 5 | 27 | 11.42 a. m. | 71 | 06 | 05 | 5ª |
| 13 | 11.45 a. m. | 71 | 06 | 22 | 5 | 29 | 11.35 a. m. | 71 | 08 | 56 | 6ª |
| 17 | . . . . . | 71 | 07 | 46 | 5 | 30 | 10.32 a. m. | 71 | 07 | 47 | 5 |
| 20 | . . . . . | 71 | 06 | 49 | 5 | 30 | 11.12 a. m. | 71 | 10 | 13 | 5ª |
| 24 | 11.15 a. m. | 71 | 09 | 32 | 5 | 30 | 11.47 a. m. | 71 | 08 | 13 | 6ª |
| 27 | 11.45 a. m. | 71 | 11 | 24 | 5 | Apr. 1 | 9.35 a. m. | 71 | 07 | 05 | 5 |
| 1889. | | | | | | 1 | 2.22 p. m. | 71 | 08 | 16 | 5 |
| Jan. 2 | 12.14 p. m. | 71 | 10 | 19 | 5 | 5 | 9.36 a. m. | 71 | 07 | 52 | 5ª |
| 2 | 3.12 p. m. | 71 | 08 | 19 | 5ª | 5 | 2.30 p. m. | 71 | 07 | 49 | 5ª |
| 2 | 4.10 p. m. | 71 | 07 | 56 | 6ª | 8 | 9.40 a. m | 71 | 09 | 32 | 6ª |
| 11 | 2.23 p. m. | 71 | 09 | 28 | 5 | 8 | 2.32 p. m. | 71 | 07 | 54 | 6ª |
| 11 | 3.10 p. m. | 71 | 07 | 05 | 6ª | 12 | 9.43 a. m. | 71 | 07 | 43 | 5 |

## Table X—Continued.

| Date. | Hour. | Dip. o | ′ | ″ | Needle. | Date. | Hour. | Dip. o | ′ | ″ | Needle. |
|---|---|---|---|---|---|---|---|---|---|---|---|
| **1889.** | | | | | | **1889.** | | | | | |
| Apr. 12 | 2.12 p. m. | 71 | 05 | 02 | 5 | July 1 | 3.10 p. m. | 71 | 07 | 44 | 5 |
| 15 | 9.45 a. m. | 71 | 07 | 35 | 5a | 5 | 9.58 a. m. | 71 | 04 | 07 | 5a |
| 15 | 2.14 p. m. | 71 | 05 | 30 | 5a | 5 | 3.05 p. m. | 71 | 00 | 41 | 5a |
| 19 | 9.50 a. m. | 71 | 08 | 20 | 6a | 8 | 9.53 a. m. | 71 | 08 | 11 | 6a |
| 19 | 2.36 p. m. | 71 | 07 | 26 | 6a | 8 | 2.35 p. m. | 71 | 05 | 56 | 6a |
| 22 | 9.57 a. m. | 71 | 06 | 34 | 5 | 12 | 10.21 a. m. | 71 | 09 | 07 | 5 |
| 22 | 2.35 p. m. | 71 | 06 | 32 | 5 | 12 | 2.57 p. m. | 71 | 06 | 07 | 5 |
| 26 | 10.25 a. m. | 71 | 07 | 56 | 5a | 15 | 11.10 a. m. | 71 | 02 | 52 | 5a |
| 26 | 12.45 p. m. | 71 | 06 | 30 | 5a | 15 | 2.55 p. m. | 71 | 02 | 52 | 5a |
| 29 | 9.35 a. m. | 71 | 07 | 54 | 6a | 19 | 9.46 a. m. | 71 | 07 | 04 | 6a |
| 29 | 2.35 p. m. | 71 | 07 | 00 | 6a | 19 | 2.41 p. m. | 71 | 07 | 00 | 6a |
| May 3 | 10.35 a. m. | 71 | 07 | 54 | 5 | 22 | 9.47 a. m. | 71 | 07 | 04 | 5 |
| 3 | 2.22 p. m. | 71 | 07 | 04 | 5 | 22 | 2.38 p. m. | 71 | 07 | 19 | 5 |
| 6 | 9.35 a. m. | 71 | 06 | 52 | 5a | 26 | 10.09 a. m. | 71 | 04 | 00 | 5a |
| 6 | 2.35 p. m. | 71 | 05 | 04 | 5a | 26 | 2.46 p. m. | 71 | 01 | 19 | 5a |
| 10 | 9.35 a. m. | 71 | 08 | 52 | 6a | 29 | 9.51 a. m. | 71 | 06 | 30 | 6a |
| 10 | 2.55 p. m. | 71 | 07 | 04 | 6a | 29 | 2.38 p. m. | 71 | 12 | 04 | 6a |
| 13 | 9.28 a. m. | 71 | 07 | 30 | 5 | Aug. 2 | 10.25 a. m. | 71 | 03 | 04 | 5 |
| 13 | 3.16 p. m. | 71 | 06 | 17 | 5 | 5 | 10.50 a. m. | 71 | 04 | 34 | 5a |
| 17 | 9.32 a. m. | 71 | 07 | 23 | 5a | 5 | 2.50 p. m. | 71 | 02 | 52 | 5a |
| 17 | 2.10 p. m. | 71 | 05 | 39 | 5a | 9 | 9.49 a. m. | 71 | 09 | 19 | 6a |
| 20 | 9.25 a. m. | 71 | 08 | 34 | 6a | 9 | 2.52 p. m. | 71 | 08 | 59 | 6a |
| 20 | 2.42 p. m. | 71 | 06 | 43 | 6a | 12 | 10.58 a. m. | 71 | 08 | 15 | 5 |
| 24 | 9.30 a. m. | 71 | 08 | 04 | 5 | 12 | 2.48 p. m. | 71 | 02 | 29 | 5 |
| 24 | 2.30 p. m. | 71 | 07 | 52 | 5 | 16 | 10.25 a. m. | 71 | 03 | 19 | 5a |
| 27 | 9.32 a. m. | 71 | 06 | 54 | 5a | 16 | 2.26 p. m. | 71 | 03 | 22 | 5a |
| 27 | 2.51 p. m. | 71 | 05 | 49 | 5a | 19 | 9.52 a. m. | 71 | 07 | 19 | 6a |
| 31 | 9.37 a. m. | 71 | 07 | 35 | 6a | 19 | 2.45 p. m. | 71 | 06 | 52 | 6a |
| 31 | 3.02 p. m. | 71 | 07 | 13 | 6a | 23 | 10.05 a. m. | 71 | 07 | 22 | 5 |
| June 4 | 9.10 a. m. | 71 | 07 | 37 | 5a | 23 | 2.27 p. m. | 71 | 04 | 19 | 5 |
| 4 | 2.32 p. m. | 71 | 05 | 31 | 5a | 26 | 10.14 a. m. | 71 | 03 | 26 | 5a |
| 7 | 9.32 a. m. | 71 | 08 | 01 | 6a | 30 | 9.50 a. m. | 71 | 06 | 41 | 6a |
| 7 | 3.00 p. m. | 71 | 06 | 35 | 6a | 30 | 2.06 p. m. | 71 | 04 | 30 | 6a |
| 10 | 9.52 a. m. | 71 | 07 | 24 | 5 | Sept. 2 | 10.31 a. m. | 71 | 06 | 11 | 5 |
| 10 | 2.35 p. m. | 71 | 02 | 50 | 5 | 2 | 2.34 p. m. | 71 | 05 | 00 | 5 |
| 14 | 10.14 a. m. | 71 | 07 | 59 | 5a | 6 | 10.06 a. m. | 71 | 04 | 29 | 5a |
| 14 | 2.41 p. m. | 71 | 06 | 41 | 5a | 6 | 2.48 p. m. | 71 | 02 | 52 | 5a |
| 17 | 9.51 a. m. | 71 | 05 | 03 | 6a | 9 | 11.10 a. m. | 71 | 04 | 22 | 6a |
| 17 | 2.10 p. m. | 71 | 03 | 24 | 6a | 9 | 2.47 p. m. | 71 | 09 | 30 | 6a |
| 21 | 10.23 a. m. | 71 | 07 | 28 | 5 | 13 | 11.09 a. m. | 71 | 09 | 19 | 5 |
| 21 | 2.40 p. m. | 71 | 06 | 52 | 5 | 13 | 2.48 p. m. | 71 | 05 | 22 | 5 |
| 24 | 10.53 a. m. | 71 | 06 | 00 | 5a | 16 | 10.54 a. m. | 71 | 03 | 30 | 5a |
| 24 | 2.42 p. m. | 71 | 04 | 52 | 5a | 16 | 2.07 p. m. | 71 | 04 | 07 | 5a |
| 28 | 9.30 a. m. | 71 | 06 | 54 | 6a | 20 | 10.31 a. m. | 71 | 09 | 49 | 6a |
| 28 | 2.27 p. m. | 71 | 07 | 52 | 6a | 20 | 1.07 p. m. | 71 | 09 | 41 | 6a |
| July 1 | 9.45 a. m. | 71 | 07 | 05 | 5 | 23 | 10.41 a. m. | 71 | 10 | 41 | 5 |

## TABLE X—Continued.

| Date. | Hour. | Dip. | | | Needle. | Date. | Hour. | Dip. | | | Needle. |
|---|---|---|---|---|---|---|---|---|---|---|---|
| 1889. | | ° | ′ | ″ | | 1889. | | ° | ′ | ″ | |
| Sept. 23 | 2.59 p. m. | 71 | 02 | 37 | 5 | Nov. 1 | 11.16 a. m. | 71 | 04 | 56 | 5ᵃ |
| 27 | 11.45 a. m. | 71 | 05 | 49 | 5ᵃ | 1 | 1.54 p. m. | 71 | 09 | 41 | 5ᵃ |
| 27 | 2.25 p. m. | 71 | 04 | 11 | 5ᵃ | 4 | 10.12 a. m. | 71 | 08 | 45 | 6ᵃ |
| 30 | 11.12 a. m. | 71 | 06 | 49 | 6ᵃ | 4 | 2.21 p. m. | 71 | 08 | 49 | 6ᵃ |
| 30 | 2.21 p. m. | 71 | 06 | 15 | 6ᵃ | 8 | 10.11 a. m. | 71 | 08 | 00 | 5 |
| Oct. 4 | 10.46 a. m. | 71 | 07 | 41 | 5 | 8 | 2.50 p. m. | 71 | 14 | 34 | 5 |
| 4 | 2.02 p. m. | 71 | 05 | 34 | 5 | 11 | 10.51 a. m. | 71 | 05 | 04 | 5ᵃ |
| 11 | 10.55 a. m. | 71 | 04 | 30 | 5ᵃ | 11 | 1.59 p. m. | 71 | 02 | 15 | 5ᵃ |
| 11 | 2.22 p. m. | 71 | 05 | 30 | 5ᵃ | 15 | 10.00 a. m. | 71 | 06 | 34 | 6ᵃ |
| 14 | 11.00 a. m. | 71 | 06 | 23 | . . | 15 | 2.27 p. m. | 71 | 08 | 41 | 6ᵃ |
| 14 | 2.52 p. m. | 71 | 04 | 19 | . . | 18 | 10.23 a. m. | 71 | 07 | 41 | 5 |
| 18 | 10.05 a. m. | 71 | 05 | 26 | . . | 18 | 2.22 p. m. | 71 | 07 | 22 | 5 |
| 18 | 2.47 p. m. | 71 | 06 | 11 | . . | 22 | 10.28 a. m. | 71 | 07 | 34 | 5ᵃ |
| 21 | 10.44 a. m. | 71 | 04 | 52 | 5ᵃ | 25 | 10.20 a. m. | 71 | 05 | 41 | 6ᵃ |
| 21 | 2.14 p. m. | 71 | 05 | 26 | 5ᵃ | 25 | 2.20 p. m. | 71 | 06 | 22 | 6ᵃ |
| 25 | 10.15 a. m. | 71 | 10 | 37 | 6ᵃ | 29 | 11.10 a. m. | 71 | 07 | 17 | 5 |
| 28 | 11.42 a. m. | 71 | 14 | 19 | 5 | 29 | 2.27 p. m. | 71 | 06 | 37 | 5 |
| 28 | 2.34 p. m. | 71 | 06 | 37 | 5 | Dec. 2 | 10.45 a. m. | 71 | 06 | 41 | 5ᵃ |
| | | | | | | 2 | 2.29 p. m. | 71 | 05 | 37 | 5ᵃ |

o